FOUNDATIONS OF LOGICO-LINGUISTICS

SYNTHESE LANGUAGE LIBRARY

TEXTS AND STUDIES IN
LINGUISTICS AND PHILOSOPHY

VOLUME 2

WILLIAM S. COOPER

FOUNDATIONS OF LOGICO-LINGUISTICS

A Unified Theory of Information, Language, and Logic

D. REIDEL PUBLISHING COMPANY

DORDRECHT : HOLLAND / BOSTON : U.S.A.

Library of Congress Cataloging in Publication Data

Cooper, William S.
Foundations of logico-linguistics.

(Synthese language library ; v. 2)
Bibliography: p.
Includes index.
1. Language and logic. 2. Information theory. I. Title.
II. Series.
P39.C68 410 78–552
ISBN-13: 978-90-277-0876-2 e-ISBN-13: 978-94-009-9820-9
DOI: 10.1007/978-94-009-9820-9

2-0876-0679-500-NC

Published by D. Reidel Publishing Company,
P.O. Box 17, Dordrecht, Holland

Sold and distributed in the U.S.A., Canada, and Mexico
by D. Reidel Publishing Company, Inc.
Lincoln Building, 160 Old Derby Street, Hingham,
Mass. 02043, U.S.A.

For friend
CLARE

FOREWORD

In 1962 a mimeographed sheet of paper fell into my possession. It had been prepared by Ernest Adams of the Philosophy Department at Berkeley as a handout for a colloquim. Headed 'SOME FALLACIES OF FORMAL LOGIC' it simply listed eleven little pieces of reasoning, all in ordinary English, and all absurd.

I still have the sheet, and quote a couple of the arguments here to give the idea.

- *If you throw switch S and switch T, the motor will start. Therefore, either if you throw switch S the motor will start, or, if you throw switch T the motor will start.*

- *It is not the case that if John passes history he will graduate. Therefore, John will pass history.*

The disconcerting thing about these inferences is, of course, that under the customary truth-functional interpretation of *and*, *or*, *not*, and *if–then*, they are *supposed* to be valid. What, if anything, is wrong?

At first I was not disturbed by the examples. Having at that time considerable personal commitment to rationality in general and formal logic in particular, I felt it my duty and found myself easily able (or so I thought) to explain away most of them. But on reflection I had to admit that my explanations had an *ad hoc* character, varying suspiciously from example to example. Moreover, I had no idea whether these were isolated oddities or pervasive problems: for all I knew there might be many more such examples where those eleven came from. Under the influence of a temporary fit of intellectual honesty I asked Adams if such was indeed the case. It was.

I then experimented with further argument forms generated more or less randomly. It began to appear that once one got beyond the simple inference patterns discussed in logic textbooks, English arguments for which the classical rules of logic failed to work right (when applied in textbook fashion) were almost as common as those for which they did. One might almost as well flip coins as use the classical logic to try to predict which English arguments would seem reasonable and which not!

My confidence in formal logic still shaken only slightly, I concluded that either the English-to-logic 'translation' rules suggested in elementary textbooks needed very substantial elaboration, or else that English conformed better to one of the well-known nonclassical systems such as *n*-valued, intuitionistic, or modal logic than to the classical two-valued system. The latter possibility seemed especially real; after all, some of the founding fathers of the nonclassical systems had expressed their dissatisfaction with the classical conditional, explicitly mentioning its faults as a motivating factor in the establishment of their systems. I found, however, that none of these systems survived the test of actual experiment. Typically, they at first raised my hopes by reversing the predictions of the classical logic for at least some of the troublesome English examples for which the latter had failed. The problem was, as further checking always seemed to show, that they reversed its predictions in just as many cases where it had succeeded. The net gain in fidelity to what seemed reasonable in English was approximately nil.

Well, there were still a number of lesser-known systems of logic to try – systems proposed in the logico-philosophical journal literature as offering conditional connectives more 'natural' in some senses that the standard conditionals. Discouragingly, none of those I examined fared much better than their standard predecessors at capturing the properties of the colloquial *if–then* and the other connectives. In fairness it must be noted that some were never intended to do so. But I was surprised to discover that others, which did seem to make some such claim, were typically based on only one or two interesting but isolated examples of English usage – hardly a massive body of evidence. Eventually I grew weary of checking out empirically unsupported systems and stopped. It seemed to me that a claim to have captured the properties of the English *if–then* was essentially a linguistic claim, hence a scientific claim calling for empirical justification. Anyone making such a claim ought to accept the burden of presenting some evidence in support of it, I thought, and not simply leave the empirical testing as an exercise to the reader.

Of course, there is not necessarily anything wrong with a system of logic which fails to conform to ordinary English usage. But it seemed nonetheless legitimate to ask: *If none of the usual systems of logic is the logic of English, what is?* To this question there appeared to be no convincing answer in either the logico-philosophical or the linguistic literature. Conceivably the question was wrongly posed; for instance, for all I knew one of the standard systems could still be the basic underlying logic of English, but with English-logic translation rules far more subtle and elaborate than had been generally supposed. But if so, the question would merely arise in a different form, namely, What are the translation rules?

The problem of how to discover the underlying logic of a natural language seemed to me then a serious and important one, and it has left me without excuse for idleness ever since. There are two general problems involved, both formidable. The first is the matter of evidence. What sorts of empirical observations are needed to lay bare a language's 'logical' structure? Is there a practical informant technique, as might be hoped on the basis of experience in conventional linguistics, or is some radically different experimental methodology indicated? The second problem, really an elaboration of the first, is that of constructing a foundational theory within which to interpret or even motivate the observations. It is recognized these days that just as scientific theories need to be tested out against observation, so one needs a prior theoretical framework (or 'paradigm') before one can draw interesting conclusions from the data or even know what data to gather. What then should be the paradigm within which to explore natural language logic?

Issues relating to these questions have of course been illuminated by many prominent thinkers, usually from the point of view of one particular discipline. Yet the problem as a whole remains elusive. The trouble seems to be that when the matter of the logical structure of natural language is examined from the perspective of any one of the existing disciplines (with the possible exception of philosophy), the problem either cannot be perceived at all or else appears already solved. Nor is it clear that if the problem as a whole were to be solved, the solution would be recognized as a solution within any presently existing tradition. (One is reminded of the man who saw a jigsaw puzzle before and again after it had been assembled. When an attempt was made to point out to him the beautiful picture which had emerged he remarked, "But what has been accomplished? There are no new pieces there at all!")

It may be that most of what is needed to construct a unified theory of language and logic is already at hand, lacking only the connecting links. Most of the pieces of the jigsaw puzzle may already be in plain sight, in other words, in which case it is high time to start fitting them together. I personally suspect this to be the case, and hope that the present work may contribute to the fitting-together process. To be specific, I believe that it is now possible to start to integrate parts of mathematical logic, descriptive linguistics, the philosophy of language and logic, automata theory, Bayesian probability theory, and certain areas of artificial intelligence research, into a coherent logico-linguistic theory of human communication.

Having admitted to unification as my ulterior motive, I must quickly beg for charity. Tolerance is needed on the part of the reader of a work concerned with unification. Since each reader is apt to be acquainted with some but not all of the disciplines involved, a certain amount of introductory material must be included for every field touched upon – to the annoyance of those already familiar with it. A related problem is that of mathematical level. The goal of unification demands clarity as to which are the primitive terms of the theory, which the definitions, and which the theorems – in other words a formal mathematical development. Such a development is bound to be too technical for some and not detailed enough for others. The compromise adopted here is to assume familiarity with elementary logic and set theory, the rudiments of probability theory, and a nodding acquaintance with descriptive linguistics. To ease the mathematical burden all formal developments are presented in the sans-serif typeface in which this sentence is set. Readers interested only in the intuitive drift of the argument may wish to skip or skim much of this material. The mathematical proofs can in any case be omitted by those willing to take the theorems on faith.

In an area as difficult as the foundations of language and logic, the only certainty is that any large new theory will turn out to be wrong in at least some points, and will eventually be superseded. Theories should therefore be stated as precisely as possible, not because their proponents are sure they are correct, but because an exact statement makes it easier for others to discover wherein they err and to improve them. It is in the spirit of this dialectic process that the present theory is set forth in rigorous mathematical dress (in the sans-serif type) as well as by looser intuitive arguments.

Colleagues have commented that Chapter 8, in which the previously developed abstract theory is applied in a case study of an actual linguistic construction (the English conditional), makes everything else coalesce and ought on all sound heuristic principles to come at the beginning. They agree, however, that logically it has to remain near the end because it treats of hypotheses which cannot even be stated precisely, let alone tested scientifically, without benefit of the preceding theoretical development. I think the advice they would have me convey to you is this: If the going gets uncomfortably abstract, don't give up until you reach Chapter 8, and if necessary skip to it.

Many friends and colleagues influenced the ideas in this book and encouraged the writing of it. While I don't suppose they would go so far as to share the blame for it, still I'd like to mention some of them. Victor Yngve of the University of Chicago was kind enough to comment on the first draft. It was Professor Yngve who first stimulated my interest in scientific language study as early as 1958, and who several years later drew to my attention the possible significance of automata theory as a vehicle for serious pragmatic language investigations. Phyllis Baxendale, then at the I.B.M. Research Laboratory in San Jose, made possible some early computer experimentation in an area of overlap between logic and linguistics. Don Swanson of the University of Chicago commented on an early draft, after having deaned into existence the favorable conditions which allowed it to be written. My indebtedness to Ernst Adams, particularly in the matter of *if–then*, should be obvious. I have also benefited from conversations on that topic with Brian Skyrms of the University of Illinois. M. E. Maron of the University of California at Berkely offered some valuable expository advice on the early chapters. Patrick Wilson of the same institution supplied specific criticisms and general encouragement in exquisitely balanced proportions. The exposition has also profited from a number of specific suggestions by J. L. Kuhns of Operating Systems, Inc. I owe much to Paul Huizinga of the University of California and Ian Carlstrom, now of Case Western Reserve University, for checking the mathematical proofs and in one instance suggesting a stronger theorem. I was assisted by the thoughtful commentary of the publisher's reviewer. I am also grateful for the extensive and very useful notes of another reviewer whose identity is unknown to me but is presumably known to him.

Berkeley, April 1977 W.S.C.

TABLE OF CONTENTS

CHAPTER 1

INTRODUCTION

1.1 Aims

This is an exploratory study whose purpose is to investigate the possibility of, and if possible lay some of the foundations for, a unified science of language and logic. Here and throughout the term 'language' is to be understood to include any code or symbol system, natural or artificial, which functions or has the potential of functioning as a medium of linguistic communication. The term 'logic' is intended to include both deductive and inductive systems of inference in any language. A 'science' is to be understood in the usual sense of a conceptual picture and accompanying mathematical formalism within which hypotheses can be expressed precisely and tested out empirically under clear rules of evidence.

The theory to be explored is technically classifiable as a system of *pragmatics*. This word has been used variously by various authors, but always as a cover term for something more inclusive than either syntax or semantics, and usually for something concerned more explicitly with the behavior of the language users. To this extent it may appropriately be applied here. The theory might also be called a *theory of information* or of *information transfer*, though it is unrelated to the well-known information theory of Hartley, Shannon, and Wiener. The role of the concept of information in the present theory is to provide a platform upon which a theory of language and a theory of logic may be erected simultaneously in an integrated fashion.

With the exception of Chapter 8, the investigation will be confined to foundational issues as opposed to the analysis of special features of particular languages. Such general problems as how the term 'language' may be defined abstractly in such a way as to serve for both a theory of language and a theory of logic, and how linguistic evidence may be brought to bear in discovering the logical structure of natural languages, will be taken up. Detailed questions about nouns, verbs, articles, quantifiers, etc., will not. It will become evident, however, that many existing proposals for the treatment of such constructions can be adapted to the present foundational system, where their descriptive adequacy may be tested out empirically under the system's rules of evidence.

1.2 Beyond Syntax

It is safe to assume that anyone reading these pages is already convinced of the importance and fascination of the theory of language, so there is no need to reemphasize that often-made point here. More discussion-worthy is the question of just what a theory of language is or should be. There are of course many kinds of language theory; we will be concerned here only with those a linguist would call 'descriptive' (addressed to questions about what languages are actually like) and 'synchronic' (pertaining to languages at particular points in time rather than to historical changes in them). Among descriptive synchronic theories attention will be further restricted to those that are scientific and mathematically formulable. And we will be concerned in particular with the development of a language theory that goes substantially beyond the syntactic level.

It is necessary to explain the sense in which the word 'syntax' will be used in these pages since it differs from some usages of the term. The expressions 'syntax' and 'the syntactic level' will serve as cover terms for all matters relating to structural well-formedness of utterances. Thus 'syntax' is for us the study of all the issues arising out of the attempt to describe the set of well-formed sentences of a language. Such issues are characterized by the empirical evidence that becomes relevant to syntactic investigations under this definition. The evidence consists of either direct or inferred judgements by the users of a language as to whether specified utterance forms constitute acceptable utterances of the language or not. For example,

John isn't here

is a well-formed sentence of English but

**Heren't John is*

is not. This use of the term 'syntax' covers more ground than the usage to which some linguists are accustomed because it includes phonetics, phonemics, and morphology as well as the larger grammatical relationships. But on the other hand it covers less ground than it is sometimes made to do in the philosophical literature. In the writings of Rudolf Carnap, for instance, axiomatically described implication and other logical relationships among sentences are regarded as part of the 'logical syntax' of language.

With this understanding of what syntactic investigations are about, it is possible to state a view probably shared in some form by many contemporary linguists and philosophers of language. It is the opinion that *the central challenge facing language theorists today is that of finding an orderly way to extend language analysis beyond the syntactic level.* (It goes without saying, of course, that not just any odd way of extending the analysis will do. The extension must go *substantially* beyond the syntactic level, it must be scientific in the sense of being both theoretically sound and empirically well-founded, and it must satisfy other criteria as well.) This general problem area – the problem of how to 'break the syntax barrier' if you will – has been and still is the linguistic riddle of the century, and it is the central linguistic issue to which this book is addressed.

1.3 BLOOMFIELD'S DILEMMA

In elaboration of the problem of extending language analysis beyond the syntactic level it is instructive to consider what Leonard Bloomfield had to say on the subject in his classic work *Language* (1933, p. 139):

> The situations which prompt people to utter speech, include every object and happening in their universe. In order to give a scientifically accurate definition of meaning for every form of language, we should have to have a scientifically accurate knowledge of everything in the speaker's world. The actual extent of human knowledge is very small, compared to this. We can define the meaning of a speech-form accurately when this meaning has to do with some matter of which we possess scientific knowledge. We can define the names of minerals, for example, in terms of chemistry and mineralogy, as when we say that the ordinary meaning of the English word *salt* is 'sodium chloride' (NaCl), and we can define the names of plants or animals by means of the technical terms of botony or zoology, but we have no precise way of defining words like *love* or *hate*, which concern situations that have not been accurately classified – and these latter are in the great majority. . . . The statement of meaning is therefore the weak point in language-study, and will remain so until human knowledge advances very far beyond its present state.

It is always dangerous to try to paraphrase a classic passage, and the historical question of precisely what Bloomfield meant is of no real concern here anyway. But one interesting interpretation which the foregoing passage *could* be given is that Bloomfield saw the language scientist as confronted by a serious dilemma. The dilemma is that the would-be analyst of a language has only two real choices: (1) to analyze the language on the syntactic level only; or (2) to analyze in addition virtually all the 'human knowledge' used by the speakers of the language in interpreting its utterances. There is no middle ground, Bloomfield apparently believed, because there is no clear stopping-place once one has started down the road from (1) to (2). But choice (2) is unthinkable, not only because of its staggering scope but also because much of the 'human knowledge' needed for it is as yet undiscovered. Bloomfield therefore felt compelled to embrace choice (1), and was followed in so doing by several generations of American and other linguists.

The ghost of Bloomfield's Dilemma still haunts us, though moderns might not describe it in quite those terms. A contemporary interpretation of the Dilemma might run like this. It has been found possible, given enough time and talent, to formulate precise and more or less accurate descriptions of the syntax of any natural language. For example, descriptions of the syntax of English are now available which, while admittedly incomplete, at least approach the goal of describing the set of all well-formed English sentences. The recursive rules contained in these descriptions are definite and precise enough so that any impartial investigator can test out their descriptive accuracy under rules of evidence that are widely recognized as valid. Thus the linguists who followed Bloomfield in seizing the first horn of the Dilemma have been reasonably successful at their restricted task.

In contrast, the success of those who have attempted to carry scientific language description beyond the syntactic level has been at best mixed. Until the mid-1960's it was a widely held prejudice that language was by its very nature amenable to precise scientific analysis on the syntactic level only, as Bloomfield implied; or at least, that everything of scientific importance about 'meaning' is somehow reflected in syntactic structure so that nothing more need be studied. Since then a more liberal attitude has prevailed, and some interesting schools of linguistic analysis have sprung up, offering ways of extending the successful syntactic apparatus into the realm of semantics. However, there is as yet no unanimity on the issue of which of these schools offers the most promise, and there have been at least some observers who see little hope of a genuine science of language emerging from any of them until their foundations have been radically strengthened through the incorporation of appropriate foundational concepts from philosophy, logic, and elsewhere.

The recent emergence of model-theoretic semantics as a serious vehicle for natural language analysis represents an important attempt to do just that. By 'model-theoretic semantics' I mean the theory of denotation and satisfaction adumbrated by Frege, developed for the classical first-order logic by Tarski, elaborated for modal and intensional logics by Kripke and others, and recently applied to certain fragments of English in a brilliant synthesis by the late Richard Montague and his associates. Patrick Suppes writes, ". . . when it comes to semantical matters, the tradition of model-theoretic semantics that originated with Frege (1875) in the nineteenth century is *the* serious intellectual tradition of semantical analysis . . . " (1974, p. 103), and I agree with Suppes that it is by far the most powerful meaning-analytic apparatus to emerge to date. However, the model-theoretic approach to language analysis is in and of itself incomplete, its most visible inadequacy being the lack of a clear empirical criterion for the descriptive accuracy of a language description formulated along model-theoretic lines. This lack was dramatized in 1970 when Suppes, admitting to some dissatisfaction with his own criteria of empirical adequacy, asked Montague publicly and pointedly to explain what he would regard as observational evidence for or against his language descriptions (1973). It is part of the tragedy of Montague's untimely death shortly thereafter that we will never know his considered answer to this question.

As it stands, model-theoretic semantics is therefore unsatisfactory for linguistic purposes because it supplies no clear rules of evidence. But it is also unsatisfactory for another, subtler reason, namely that it assumes that the goal of semantic analysis is the statement of truth-conditions ('rules of satisfaction', etc.) for the sentences of the object language. But what if there are natural-language sentences which have no truth-conditons? This possibility has worried many philosophers of language, and rightly so. The status which should be accorded the idea of truth-conditions, it seems to me, is that of an empirical *hypothesis*, namely for any given language the hypothesis that all of its sentences have truth-conditions. The statement of truth-conditions is not a goal to be settled upon arbitrarily before the fact, but a way of formulating a conceptual picture which may or may not be an accurate picture of particular languages. Not only should the hypothesis of clear truth-conditions for every sentence not be assumed *a priori*; there is now a certain amount of linguistic evidence (to be reviewed in Chapter 8) that for at least one important natural language (English) and under plausible rules of evidence, the hypothesis should probably be rejected. In other words, there are probably at least some natural-language sentences for which it is simply inappropriate to attempt to state truth-conditions using the standard model-theoretic apparatus, sentences which are nonetheless analyzable under a broader set of assumptions which in effect contradict the model-theorist's basic postulate.

What is needed then, in my opinion, is a scientific framework within which model-theoretic semantics, and other approaches to meaning analysis as well, can be tested out and, where valid, applied. It will be a major objective of the present study to provide such a framework and to equip it with rules of evidence no less definite and precise than those to which linguists have already become accustomed in carrying out purely syntactic investigations. The proposed scientific framework is intended as a resolution of Bloomfield's Dilemma in the sense that it goes substantially beyond syntax (and for that matter beyond model-theoretic semantics too), yet does not commit the investigator to the analysis of every language-related phenomenon at once. Not circumscribed by any particular structural approach, it has, I believe, the virtue of passing in some sense 'half-way' between the horns of the beast, maximizing as it were the immediate interest and practicality of the analyses possible within its scope while minimizing the danger of bogging down in the still relatively uncharted linguistic territory that might be called 'higher pragmatics'.

1.4 THE RESEARCH STRATEGY OF THE ISOLABLE SUBSYSTEM

Let us forget linguistic history for a moment and think of the language user simply as a black box, or at any rate as a box (it may not be entirely 'black'). The input to this box consists of sensory data; its output is motor actions. Among the inputs and outputs are a special kind called 'linguistic utterances'. Now imagine a scientist who wishes to study this box with a view to finding out what he can about what goes on when the utterances are transmitted and received. Two ways in which he might attack the problem come immediately to mind. The first would involve studying only the form of the linguistic utterances themselves as they are observed going into and coming out of the box. The second would be to make the study inclusive by analyzing in addition everything that goes on inside the box relating to the processing of the utterances.

But to think of these as the only two possibilities would be to get tossed once again between the horns of Bloomfield's Dilemma. The first approach would confine the investigations to the syntactic level, while the second would be to attempt to analyze all of the user's knowledge, social habits, etc. that have any bearing on how he uses his language. Once again it is easy to see that the first approach is practical but does not go as far as one could wish, while the second is hopeless because of the complexity of the box.

What is to be done? Our hypothetical scientist could well take a lesson from the systems analysts, whose reaction would be this: When a system is too complex to analyze as a whole, one must look for an isolable subsystem of it and begin by analyzing that. With this strategy one may eventually divide and conquer; without it one might as well give up. When this 'strategy of the isolable subsystem' is adopted, all pretense of investigating simultaneously every aspect of linguistic behavior must be given up and effort focused instead (at least at first) on the analysis of whatever component of verbal behavior seems most critical and most readily isolable.

This strategy entails a radical departure from the usual methods of syntactic analysis. To mention only the most obvious distinctions, the subsystem to be analyzed will not be just the set of well-formed sentences possibly accompanied by deep structures. Instead, it will be a self-contained behavioral system with input and output and internal processing. And since it will foreseeably have at least some of the attributes of a black box, it will be necessary to analyze the internal processing by indirect techniques appropriate to a nonobservable subject matter. In view of these crucial differences it seems unlikely that any easy extension of established syntactic methods will suffice for the analysis. An entirely different analytic vehicle is needed; automata theory suggests itself as one possibility if used in its full power (as opposed e.g. to the uses to which it has been put in the theory of syntax).

The choice of subsystem on which to focus attention is obviously of critical importance to the success of this strategy. It must be inclusive enough to be of deep interest but not so inclusive as to be unmanageable. It must be clearly separable, with its own private input and output readily observable even if only indirectly so. The rules it embodies should be genuine language rules in the sense of describing the verbal behavior common to all users of the language. If in addition the analysis of the subsystem should turn out to be useful in computer applications that involve language processing, that would be a welcome side-benefit. This is the strategy that will be pursued in the sequel; an isolable subsystem with these properties will be sought out, which will then serve as the focus of the investigation.

1.5 THEORIES OF LANGUAGE VS. LANGUAGE ANALYSIS

We have used the terms 'theory of language' and 'language analysis' loosely here, and do not know how to define them exactly. Nevertheless, as a rough characterization of how they are to be understood in this book, it might be said that language analysis is the activity of writing language descriptions, while the theory of language is what tells one how to carry out that activity; i.e. a theory of language is a conceptual framework within which language analyses can be carried out. The formulation of precise, detailed descriptions of languages or language fragments is the characteristic activity of descriptive linguistics, so much so that if a theory does not bear somehow on this activity it is doubtful whether it should be classed as a general theory of language at all.

Language descriptions are important for at least two reasons. The first reason is the obvious practical one that applications of descriptive linguistics such as language teaching or language processing by computer require specific descriptions of specific language phenomena. The second and perhaps more fundamental reason is that the making of language descriptions is the main point of contact between language theory and reality. By insisting that a theory of language tell how to make detailed language descriptions, one ensures that the theory has a proper regard for the place of observation and evidence in the scientific process, as well as an adequate conceptual apparatus within which meaningful empirical hypotheses can be framed. It is the requirement that a theory get one all the way from observation of linguistic behavior to a symbolic language description that keeps the theory honest. This does not mean that a theory of language must embody a discovery procedure, so that the writing of the language description will be automatic. However, the theory must at least specify the general form that a language description must take, and provides rules of evidence for testing whether a given description having this general form is indeed an accurate description of the language it is supposed to describe. If it does not do this, it is not a theory of language in our sense, or at any rate not a mature one.

Another way to put it is that a theory of language should be both 'global' and 'particularizable'. A good general theory of language should say something about all possible languages, and in this way illuminate the nature of language in general; but at the same time it should offer a viable framework for the analysis of any particular language, providing as an integral part of the theory the essential technical apparatus necessary to formulate a full description of that language and to validate the description's accuracy empirically. A theory that meets these requirements has a dual character, offering a language model that is at once abstract and applicable to concrete cases.

The prevalent syntactic theories of language have this dual character. The simplest example of such a theory is the one often encountered in automata theory and mathematical linguistics according to which a language is represented as a set of finite sequences of elements of some finite alphabet or vocabulary, the intended interpretation being that the sequences in the set represent the well-formed sentences of the language. (Most current syntactic theories are elaborations of this simple notion, introducing such additional features as the distinction between the 'deep' and 'surface' structures of sentences.) This theory is global because it is broad enough to characterize the syntax of any language, and particularizable because there are well-known informant techniques for testing whether the set of sequences specified by a given generative language description is an accurate representation of a given language. On the other hand, the broader theories of language that take up matters outside the realm of syntax often do not have this character, it being the aim of the present work to develop one that does.

When a theory of language is both global and particularizable, there is no confusion as to whether it is intended to apply to natural or to artificial languages. If truly global, it must apply to both. The difference in the way it applies is that in the case of a natural language, the formulation of a particularized language description within the theory is a scientific enterprise involving all the usual scientific processes of observation, generalization, and inference. By contrast, the particularized description of an artificially constructed or 'formal' language such as those invented by mathematicians, philosophers, and logicians can be discovered directly simply by interrogating its creators, who are after all the only ultimate authority. This is why studies of natural languages have an empirical, and artificial languages a deductive, flavor.

1.6 Theories of Logic

Webster defines 'logic' to be "a science that deals with the canons and criteria of validity of inference and demonstration". Note especially the definition's characterization of logic as a *science*, a characterization which we will take seriously in a sense to be explained presently. Generally speaking, we will employ the term 'logic' in a manner consistent with the foregoing definition, with the difference that we will put no special emphasis on 'inference and demonstration' over-and-above equivalence, inconsistency, tautology, or any of the other well-known logical properties and relationships that can hold among sentences or sets of sentences. For us, logic will consist of the study of the entire family of possible logical relationships.

Now if logic is the study of logical interrelationships, logical theory must presumably be the place in which to find explications of these interrelationships. Thus, *the first responsibility of a theory of logic is to provide general characterizations of logical implication and the other logical relationships.* It might do more, but if it does not do at least this much it is hardly a general theory of logic in any very pregnant sense of that phrase. In other words, a general theory of logic should supply acceptable abstract definitions, precisely formulated, for what it means for one sentence to imply another, to be inconsistent with another, and so forth.

But to be fully general a characterization of a logical relationship must apply across all possible languages. Most textbooks and other works on logic make no attempt to state precisely a general theory of logic in this sense. What they do instead is to define the logical relationships for a few particular formalized languages, often just one well-known language such as the classical predicate calculus; and though the definitions are commonly given in detail and lack nothing in rigor, they are nonetheless specialized to particular languages. There is nothing wrong with this approach so long as it is not claimed to describe the intrinsic nature of the logical relationships in general, but particularized treatments fail to provide a comprehensive theory of logic of the kind envisioned here.

A few recent textbooks do attempt to provide abstract characterizations of the logical relationships applicable across all languages. Typically these are model-theoretic or truth-conditional characterizations. As already discussed, one cannot quarrel with these semantical definitions when they are applied to formal languages tailored to meet their semantical requirements, but one must challenge the assumption that they apply to all possible languages. Some languages are semantically structuralizable and some, probably including the natural languages, are not, as we shall see. Hence a more general characterization of the logical relationships is needed than the standard model-theoretic apparatus provides.

It was said earlier that a good theory of language should be both global and particularizable, and now the same can be seen to hold true of a good theory of logic. Its fundamental concepts should apply in their general statement to all languages, but should be potentially describable in particular detail for any given language. In a theory of logic of this kind, general questions as to e.g. the nature of implication can be discussed in a language-independent manner without sacrificing the possibility of specifying that relationship accurately for any specific language. It is another major goal of this study to provide a theory of logic fulfilling this desideratum. This logical goal does not conflict with, and in fact will be seen to complement, the linguistic aim of carrying language analysis substantially beyond the syntactic level.

Here again the question of natural versus artificial languages arises. The logical relationships have to date been specified in precise detail mainly for the artificial languages studied by mathematical logicians. This historical circumstance may have tended to create the mistaken impression that the study of logical relationships among sentences of natural languages does not properly fall within the province of logic. But in reality it does, for a global theory of logic applies by definition to *all* languages, formal and natural.

This line of thought leads to the acceptance of the idea of 'natural logic' proposed by some linguists as the counterpart in the natural languages to the established tradition of logical and metamathematical investigation in the formal languages. The possibility of a science of 'natural logic' is an exciting one with which we shall be implicitly concerned throughout the sequel. It holds out the prospect of an eventual shift of emphasis in logic away from the traditional 'prescriptive logic' toward 'descriptive logic' – a shift which, if it comes about, would be analogous to the shift of interest among linguists from prescriptive to descriptive grammar that took place a half-century ago.

The crucial difference between natural logic and the logic of the formal languages is that the former is a scientific enterprise involving the description of natural phenomena while the latter is mainly mathematical and philosophical. The logic of a natural language cannot be discovered simply by asking the language's inventors what its rules are. The Creator of the natural languages is, inconveniently, not readily accessible for this type of interrogation. The investigator of natural logic must therefore resort to observation, experimentation, inductive generalization, and all the other trappings of empirical science. As a consequence, his activities are apt to be radically different from those of the traditional logician even though both may subscribe to the same general theory of logic, for the one is a discoverer and the other an inventor.

Practically speaking, natural logic is apt to entail a great deal of hard work, occasioned by the intricacy of the natural languages. The promise of reward is correspondingly high, however. It would be foolhardy to assume that millenia of social evolution have left in the natural languages no important clues to the nature of human reason.

1.7 Logico-Linguistics

Thinkers have long been convinced of an intimate connection between language and logic. Most of the great philosophers of language have been philosophers of logic as well, and this is in itself evidence of the connection. Yet the exact nature of the link remains controversial. One way to clarify it would be to develop a theory of language, then a separate theory of logic, and then a theory relating them. Ideally, however, *the foundations of language and logic should be developed simultaneously as a unified theory*, for if this can be done the relationship of the two will be self-evident. There is no guarantee ahead of time that this is possible, but it is a desideratum.

The theory to be examined in these pages attempts to satisfy this desideratum along with the others that have been mentioned, taking on the appearance of either a theory of language or a theory of logic depending on the perspective from which it is viewed. Since the unifying concept is the notion of information storage and transfer, it is also an information theory, but what it says about information *per se* is probably of less interest than the way the concept is related to the notion of a language or a logic. By pursuing the idea of information in one direction a theory of language is arrived at, while following up its other implications yields a theory of logic based on the same conceptual underpinnings.

Becuase the theory extends well beyond the syntactic level and deals specifically with the language users in relationship to the information they transmit, we will refer to it occasionally as a 'pragmatic' theory. We will also call it a 'logico-linguistic' theory because of its dual character as both a theory of language and a theory of logic. My own use of the term 'logico-linguistics' goes back to an internal report written in 1961. It has also come into occasional use among some philosophers and linguists in connection with the application of certain ideas and techniques of logic to problems of language analysis. Here the word will be used in the broadest sense of a unified science sharing the aims of both linguistics and logic.

CHAPTER 2

INFORMATION AND LANGUAGE

In this chapter the strategy of the isolable subsystem is followed out in an attempt to find a body of verbal behavior suitable for linguistic analysis beyond the syntactic level. The result, the 'Language Automaton', suggests a pragmatic definition of language. The starting point is the concept of an 'information state'.

2.1 INFORMATION STATES

The most important primitive notion of the theory we wish to consider is that of a state of informedness or *information state*. Because it is a primitive term no precise definition of it can be stated, but some motivating remarks can be made.

What is wanted for the study of information is a conceptual construct of some kind that embodies only the most fundamental objects and operations encountered in dealing with information. Probably the most basic of all notions of this sort is simply the idea of information existing somewhere – that is, of information residing at a certain time in some particular location. This concept of 'stored' information implies the existence of a system in which the information is stored, the system being a physical complex of some sort whose organization or behavior is indicative of the information it embodies. It is this way of thinking about information that gives rise to such locutions as "The information will be available at such-and-such a time and in such-and-such a form."

There may be other legitimate ways of conceiving information, for example information as it exists in the act of being processed, transformed, or communicated. It seems likely though that these dynamic concepts of information, if definable at all, would be dependent upon the static notion as a prior concept. To process information is to change the form in which it is stored.

Examples of systems often considered to contain stored information include: (a) a library; (b) a hologram; (c) a DNA molecule (conceived as a store of hereditary information); (d) the internal memory of a digital computer; and (e) the human brain. (In example (e), the person of whom the information-storing system is a part is said to 'know' or 'believe' the information in question.) Now in order to gain a clear understanding of what stored information is, we must discover what it is that such systems have in common. But what in the world do (a)-(e) have in common?

It would seem that the possible ways of storing information are so diverse that *nothing* of general applicability can be said about the physical structures involved, or at any rate nothing helpful. If the goal is to find a property general enough to characterize simultaneously all information-bearing structures, the quest seems hopeless. What is needed for a general theory of information, then, is a formulation of the notion of stored information so abstract that it assumes nothing about the particular form the information takes. The only concept that would appear abstract enough is the fundamental notion of a 'state'.

In systems theory the set of possible states a system could be in is called its *state set* and the state it is actually in at a given moment its *current state*. Applying these ideas to examples (a)-(e), one finds in each case there is indeed a physical organization potentially capable of being in any of a large number of states, though actually in only one at any given time. One sees too that the state of the system at a given time is what determines the information it is conceived to contain at that time. Knowing its current state is the same as knowing what information is currently stored in it.

To deal abstractly with the concept of stored information we need therefore make only the following frugal assumptions: (i) there is a physical system in which the information is stored; (ii) there is a set of states which the system has the potential of being in; and (iii) at any given time (except possibly during temporary transitional periods) the system is in one and only one of these states. It is assumed that the state set can be specified in such a way that knowledge of the current state is sufficient for determining all the information the system currently contains. We will speak of an *information state set* and an *information state* when referring to state sets and states in this connection.

2.2 Input and Output

What further concepts are needed for the study of information? Well, a capacity for storing information is of no use unless there is some way of getting the information into and out of storage. This suggests that the idea of an information-containing system should be supplemented by the auxiliary notions of input and output.

Let us consider input first. To put information into a system is to set the system into a desired information state. For some systems it may accordingly be easiest to regard an 'input' directly as a state-changing operation, while for others it may be more convenient to identify inputs with objects or messages whose submission causes such operations. In any case the search for structural properties common to all inputs seems as hopeless as the search for structural properties common to all states was, and so we assume for the general case only that there is indeed a non-empty set of possible inputs associated with every information-storing system. Let us call them *informative inputs*.

By way of illustration imagine a computerized weather information system designed to keep track of current weather conditions around the globe. In computer memory is what amounts to a symbolic weather map of some kind which is changing constantly as new weather data come in. This is an information-storing system whose information state set is the set of all weather maps representable within system memory. A typical infomative input might be a message such as *London: 61°* or *New York City: Fair*. The informative input set would be the set of all such input messages to which the system is programmed to respond by making appropriate changes in the internal map. The state change brought about by even a single input may be far-reaching, for in an advanced system the input message might allow inferences to be made about the weather in parts of the world other than the one mentioned explicitly in the message.

The matter of input and of information state changes resulting from input bears on the entire problem of what it means to 'inform'. If the information state of a system is regarded as representing a fund of information, incoming meassages that change the state may be regarded as adding to, updating, or in some other way changing this fund. There is a sense in which the system is indeed being 'informed'. And since informing and being informed are what communication is all about, the study of input signals and the state changes they bring about opens the theory of information states up into a theory of communication also.

Next let us consider how information may be gotten out of an information-storing system. Obviously, output of some sort is essential if the system is to be more than an information sink. The output may take any of a number of forms, however; for example it may be generated spontaneously or it may be elicited in response to special 'test inputs'. For the sake of definiteness we will consider the latter arrangement; more specifically we will assume there are only two possible outputs interpretable as 'Yes' or 'No' responses to test inputs. A *test input*, or member of the *test input set*, may be associated with possible properties of information states; when a test input is received the yes-or-no output generated in response tells whether the current information state possesses the associated property or not. This output arrangement, which for conceptual purposes we will arbitrarily assume to be in force, has the virtue that it allows in principle for information of any kind to be elicited from the information-storing system by submitting appropriate test inputs. Other output arrangements are always replaceable in principle by one of this form.

Illustrating with the computerized weather system again, the test inputs might be such messages as *New York City: Fair? Paris: Fog?* etc. A 'Yes' output in response to one of these queries would indicate that the current internal weather map did indeed have the property in question, for example the property of recording fair weather in New York City. Some types of information could be elicited from such a system only with awkwardness – to find out the temperature in London one would have to submit a long series of test inputs such as *London: 30°?*; *London: 31°?*; *London: 32°?*; etc. Nevertheless the desired information is accessible in principle. Note well that a 'No' answer to *Paris: Fog?* does not mean that the system contains the information that there is no fog in Paris. It merely indicates that no information is stored to the effect that there is fog. There might, for example, be no information at all currently stored in the system about weather conditions in Paris.

The study of such outputs is important, among other reasons, for the interrelationships among the test inputs which they manifest. For example, if one were to ascertain that no matter what the internal weather map a 'Yes' response was forthcoming for *London: Precipitation?* whenever *London: Rain?* produced a 'Yes', one might be led to hypothesize the existence of a special (logico-)linguistic relationship between these two messages.

To the assumptions of the last section we have now added the following: (iv) an information-storing system has an associated set of possible informative inputs, the receipt of a member of this set causing in general a change of information state; and (v) it also has a set of possible test inputs to any of which it will respond with a binary output dependent on the current information state. The stipulation that the output be binary is inessential and will be relaxed later on. Apart from variations in output arrangements, it would be hard to think of any information-containing devices or 'knowing systems' which would be excluded by requirements (i)-(v).

2.3 INFORMATION AUTOMATA

As a next step it seems natural to look for a way of bringing the concepts considered so far together in the form of a single abstract entity. The entity which achieves this is what we will call an 'information automaton' – a device-type exemplified by our weather information system.

The form of an information automaton is indicated in the black-box diagram of Figure 2.1. The box has an informative input set, the members of which (upper input arrow) may or may not have the physical form of symbolic messages. An informative input causes no output but causes in general a change in the internal state of the device. The device is also equipped to receive test inputs (lower input arrow) from a test input set. A test input causes no change of state but gives rise to a binary output interpretable as a clue about the nature of the current state. It is assumed for convenience that no two inputs ever reach the device simultaneously, for example a test input is never received at the same time as an informative input. The whole diagram is readily translated into mathematical terminology, which is its only purpose: it is merely a way of visualizing an abstract mathematical entity.

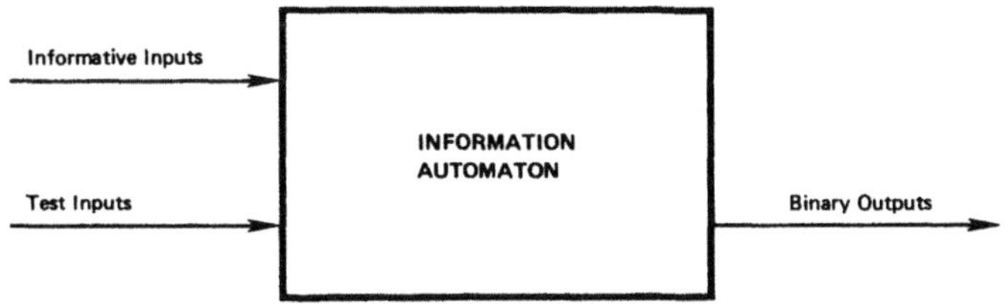

Figure 2.1 Conceptual diagram of an information automaton.

We will adapt classical set theory as a metalanguage in which to give precise expression to the notion of an information automaton. So strong a set theory may be unnecessary, and an adequate constructivist or intuitionist underpinning for the theory could probably be worked out; but the convenience and familiarity of garden-variety set theory recommends it for initial attempts at stating foundations for an empirical science. Using set theory, the notion of an information automaton is readily formalized by drawing on automata theory. Automata theory, the branch of systems theory treating of systems with discrete input–output, recognizes as an important class of systems those that are 'sequential' in the sense that each input gives rise to one and only one output, and 'deterministic' in the sense that when a given input is received in a given state, the ensuing state change and output are completely determined.

The best-known mathematical model for a sequential deterministic automaton is the so-called 'Mealy model'. Adopting the customary convention according to which an abstract system is represented by an n-tuple of the entities characterizing it, the Mealy model automaton is defined as follows.

DEFINITION 2.1. For any $\mathfrak{A}$, $\mathfrak{A}$ is an *automaton* if and only if $\mathfrak{A}$ is a 5-tuple $\langle X, Y, Z, \delta, \lambda \rangle$ in which:

(i) X is a non-empty set (the 'input set');

(ii) Y is a non-empty set (the 'output set');

(iii) Z is a non-empty set (the 'state set');

(iv) δ is a (partial) function (the 'state transition function') from $Z \times X$ into Z;

(v) λ is a (partial) function (the 'output function') from $Z \times X$ into Y.

This definition is to be found in standard reference works on automata theory (e.g. Starke 1969, p. 22). Note the absence of finiteness requirements. Because such systems may not be physically realizable they are sometimes called 'abstract automata'.

(The term 'partial' in the definition brings up a minor technical complication that will recur parenthetically throughout the sequel. A *total function* assigns to every element in its domain a unique element in its counterdomain. In contrast, a *partial function* assigns a unique element of its counterdomain to some but not necessarily all elements of its domain. A partial function is said to be *defined* for an element of its domain if and only if the element is one of those to which the function assigns a unique member of the counterdomain. In an automaton the state transition function δ and output function λ are partial functions, defined in general only for certain pairs in $Z \times X$. This provision is needed because in some automata the state transition and output rules are not well-defined for certain state-input pairs, a situation which could arise if for example there are certain states in which certain inputs are never expected to be received. Automata with partial state transition and output functions are called *incomplete automata* (see e.g. Ginsberg 1962). In case δ is total the automaton is said to be *δ-complete*, and when λ is total it is *λ-complete*. An automaton which is both δ-complete and λ-complete is *complete*.)

Information automata are defined by adding appropriate restrictions to the general definition of an automaton. Suppose X_I and X_T are the informative and test input sets of an information automaton. Since these two sets could overlap, their members must in general be tagged in some way to indicate whether they are being used as informative or test inputs. Let us adopt the convention that an input followed by a period is an informative input while an input followed by a question mark is a test input. Formally an informative input is an ordered pair $\langle x, .\rangle$ with $x \in X_I$ and a test input has the form $\langle x, ?\rangle$ with $x \in X_T$. Let the possible output signals be denoted by '**Y**' and '**N**' interpreted as yes-or-no responses. When an input results in no output at all, the output is said to be the null output Λ.

DEFINITION 2.2. $\mathfrak{A}$ is an *information automaton* if and only if $\mathfrak{A}$ is an automaton $\langle X, Y, Z, \delta, \lambda\rangle$ and there exists a set X_I ($\mathfrak{A}$'s 'informative input set') and a set X_T (its 'test input set') such that

(i) $X = (X_I \times \{.\}) \cup (X_T \times \{?\})$;

(ii) $Y = \{\mathbf{Y}, \mathbf{N}, \Lambda\}$;

(iii) for every $z \in Z, x \in X_T$,

$$\delta(z, \langle x, ?\rangle) = z;$$

(iv) for every $z \in Z, x \in X_I$,

$$\lambda(z, \langle x, .\rangle) = \Lambda.$$

and for every $z \in Z, x \in X_T$,

$$\lambda(z, \langle x, ?\rangle) \neq \Lambda.$$

Clause (iii) of the definition says test inputs never cause a change of state and (iv) says null outputs are given in response to informative inputs and informative inputs only.

(Information automata need not be complete. It is understood that $\delta(z, \langle x, ?\rangle)$ and $\lambda(z, \langle x, .\rangle)$ are always well-defined with values as indicated in (iii) and (iv). However $\delta(z, \langle x, .\rangle)$ and $\lambda(z, \langle x, ?\rangle)$ are not necessarily well-defined for all z and x.)

Many features of the formal definition are arbitrary, and other conventions would have captured the intuitive concept of an information automaton just as well. For instance, readers acquainted with automata theory will recognize that the Moore model automaton could have been used instead of the Mealy as the basis of the definition. This could have been done by taking information automaton outputs to be sets of messages instead of binary signals and doing away with test inputs entirely.

2.4 LANGUAGE AUTOMATA

The concept of an information automaton is broad enough to take in virtually all situations involving the storage and accession of information, subject only to the very general qualifications of sequentiality (one-at-a-time input and output) and determinacy. We will be interested for the most part in a narrower class of systems than this, however. Information automata will be of special interest when their input is 'linguistic', consisting of symbolic messages of some kind. An information automaton whose informative and test input sets are the same class of symbolic utterance types will be called a ***language automaton***. Language automata will constitute the main subject matter of the sequel.

Implicit in the idea of a language automaton is the assumption that in most languages of interest the messages are representable as finite strings of symbols. The symbols are understood to be linguistic units of some kind, and the symbol strings are linear arrangements of these units through time or space. The linguistic units could be phonemes or morphemes, or in the case of a written language graphemes or lexemes, and the possibility of larger units such as words or even phrases is not ruled out. The symbol strings as wholes represent 'sentences' of the language which are meaningful in isolation. The alphabet or vocabulary of basic linguistic units is finite, but the set of sentences constructable from them may be either finite (as in some simple signal systems) or infinite (as in most artificial and all natural languages). We restrict attention to languages whose messages are amenable to representation in this standard fashion. The class of messages serving as a language automaton's informative and test input set will be called its *sentence set*.

DEFINITION 2.3. $\mathfrak{A}$ is a *language automaton* if and only if $\mathfrak{A}$ is an information automaton with informative input set X_I and test input set X_T and there exists a set S (the 'sentence set' of $\mathfrak{A}$) such that for some finite set V (the 'vocabulary' of $\mathfrak{A}$), S is a non-empty set of finite non-null sequences of members of V and $X_I = X_T = S$.

When a language is of interest, its sentences may be associated with the sentence set of an abstract language automaton. The idea is to associate particular language automata with particular languages in such a way that the automata reflect the properties of the languages as communication systems. In this way language automata may be used as surrogates for languages in the investigation of linguistic information transfer. For example a language automaton for English would be the object of interest in a study of communication in English.

Although the term 'language' has not yet been defined precisely, it may be understood roughly to denote any system of symbol-sequences (sentences) with accompanying interpretive conventions making them useable for the purpose of conveying information. In this sense of the term, which seems close to the colloquial one, many insect and animal signal systems would be languages as would most gesture, semaphore, and other signal systems used by humans. The 'formal languages' used by logicians and mathematicians, provided they are applied calculi with at least a potential capacity for conveying factual information, are also languages in this sense, and so of course are all historically existing natural languages. Formal calculi useable only as aids to private deduction or computation, or only for conveying 'logical' information as opposed to factual, will be regarded as languages of a degenerate type to be discussed later under the rubric of 'uninformative' languages. Excluded entirely from the class of languages to be considered here are the so-called 'private languages' sometimes discussed by philosophers, as well as any languages capable of conveying only emotion or poetic imagery.

To simplify the discussion we will postpone consideration of some of the troublesome features found in many languages, especially the natural languages. Interrogative, imperative, exclamatory, and other nondeclarative sentence types will be ignored for the time being, as will vague, ambiguous, and context-dependent sentences.

With these understandings in force the idea of associating languages with language automata comes naturally. A language automaton for a language is one which acts, as it were, as though it 'understood' the sentences of the language. That is to say, it reflects the characteristics of the language in the following ways: (1) the sentence set of the automaton must represent the set of all well-formed sentences of the language; (2) the state set of the automaton must contain representations for all possible funds of information falling within the expressive range of the language; (3) sentences 'learned' as informative input must cause state changes appropriate to what the sentences say; and (4) sentences must be 'believed' or not (i.e. give 'Yes' or 'No' responses) according as the information expressed in them is or is not included in the current fund. These conditions simply reflect the motivating ideas behind all information automata, specialized only to the extent that the informative and test inputs are now regarded as the sentences of some definite language.

EXAMPLE 2.1. *The Detectives' Language.* Imagine a little linguistic community composed of a group of detectives working on a murder case. The detectives are scurrying about the house in which the murder took place, gathering evidence and occasionally telling each other their conclusions. There are two principal suspects, Coe and Doe, and the burning issue is: Which of the two (if either) was in the house on the night of the murder? So intent are the detectives on this question that the language in which they are communicating with one another is the fragment of English consisting of just the four sentences *Coe was in, Doe was in, Coe was out*, and *Doe was out.* All four are declarative and unambiguous, and their meaning is clear to the members of the linguistic community in question – the detectives all understand that the temporal reference is to the time of the murder, for instance.

To associate a language automaton with this four-sentence language, the sentence set is taken to be the set containing the four sentences. The state set must contain a state for each different fund of information any detective might have on the subject of Coe's or Doe's whereabouts. The state transition function must have the property that when a sentence is received as an informative input (i.e. when a detective hears a colleague utter the sentence and believes him) the new state that is entered contains the new information just received in addition to all prior information consistent with it. And the output function must reflect accurately the information associated with the various states.

A language automaton fulfilling these requirements is displayed in Figure 2.2 by means of a state diagram. It is standard practice in automata theory to use state diagrams as an informal means of representing simple automata, and the diagram of Figure 2.2 is standard except for simplifications made possible by the specialization to language automata. The circles represent states, a sentence is written within a circle when a 'Yes' answer would be given to the sentence if received as a test input in that state, and an arrow indicates the new state entered when the sentence written along the arrow is received as an informative input.

State z_1 in the middle of the diagram represents a state of total ignorance about Coe's and Doe's whereabouts such as one might expect to find in a detective who had just come on the scene. Because no sentence is assertable in this state (technically, because all four would yield a 'No' output), nothing

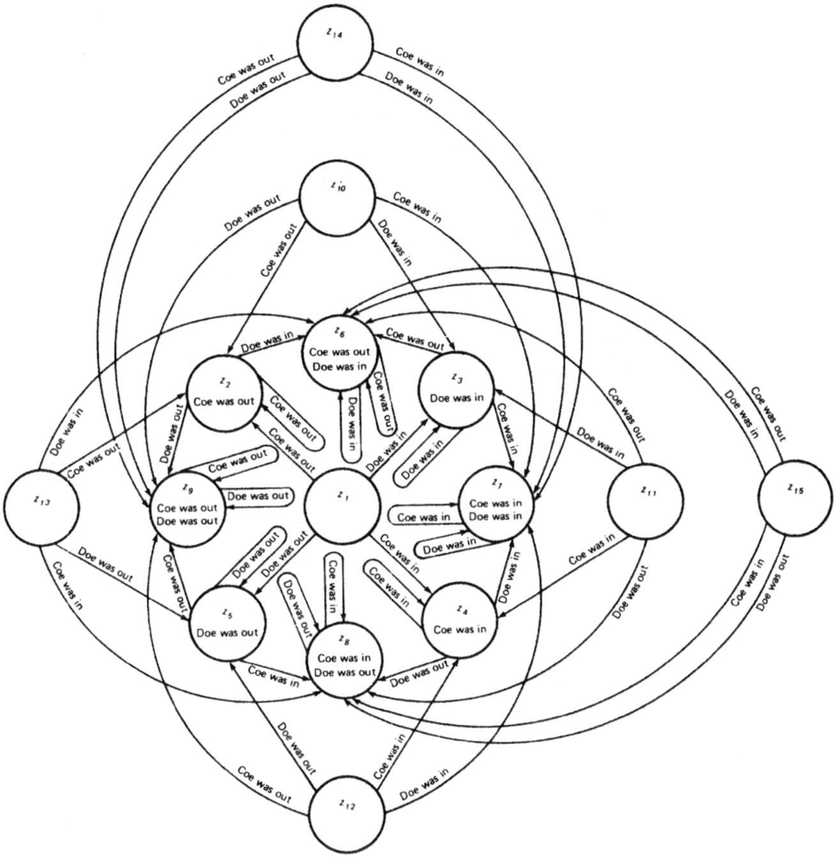

Figure 2.2. State diagram of a language automaton for the Detectives' language.

is written inside z_1's circle. But if, say, *Coe was out* is heard by a detective in state z_1, he will undergo a change of state to z_2, as the arrow indicates.

States z_{10} through z_{15} are of special interest. They are like z_1 in that they represent states in which no sentence can be asserted, but they differ from z_1 in that they represent conditional information rather than no information at all. A detective in z_{10}, for instance, is convinced that *if* Coe was in Doe must have been in, too. This belief is evidenced by the arrow from z_{10} to z_7, a state in which both *Coe was in* and *Doe was in* are known.

There are seven such states in which none of the four sentences is believed. There is a special lesson in this. *An information state cannot always be uniquely determined by the set of sentences believed in that state.* Were it otherwise one could dispense with automata theory and deal just with sets of sentences; but the example shows that sets of sentences are not always adequate as representations of belief states. One simply cannot count on the sentences of a language having sufficient expressive power to specify all the information states needed to analyze the language, except for special languages in which this is known ahead of time to be the case. Here is one reason why it is necessary to bring to bear in language theory the full apparatus of automata theory or something of comparable analytic power.

Figure 2.2 is intended only as an illustration of the general kind of analysis made possible by automata theory; it is not claimed to be necessarily the most accurate possible analysis of the Detectives' language. One of its drawbacks is that it makes no allowances for belief reversal. Once a detective has decided, for example, that Coe was out, according to the diagram he will (barring new nonliguistic evidence) spend the rest of his life in states that believe Coe was out, even if told later on highest authority that Coe was in. (The diagram shows no arrows for, and hence leaves undefined, state changes that would occur if *Coe was in* were learned while believing *Coe was out.*) This unnatural feature of the analysis reflects a property called 'intransigence' to be discussed later.

When a sentence is received as an informative input by a language automaton, we will speak of the ensuing state change as a manifestation of the automaton's *learning operation*, and the automaton will be said to 'learn' sentences submitted to it in the informative input mode. This is of course not the usual trial-and-error learning studied in stimulus–response psychlogy, but the term 'learning' seems appropriate nonetheless; for even though there is only one stimulus – the sentence – still there is a change of information state. Some commentators have expressed dissatisfaction with what they see as a too exclusive concern among learning theorists with many-trial learning, and cite speech recognition as an obviously significant case of virtually instantaneous learning. Linguistic communication is single-trial learning *par excellence*.

Technically a language automaton's learning operation is a two-place function $L\colon Z \times S \rightarrow Z$ defined by the identity $L(z, s) = \delta(z, \langle s, .\rangle)$. An expression of the form '$L(z, s)$' may be read "the new state entered upon receiving s as informative input while in state z", or more briefly "the result of learning s in z".

As a companion term a language automaton's responses to test inputs may be said to be governed by its *belief relation*. When a test sentence produces the 'Yes' response, it means that language users in the automaton's current state would know or believe the sentence. 'Belief' is best regarded here as a purely technical term: the belief relation holds between a state and a sentence just in case the sentence submitted as a test input would produce a 'Yes' response in that state. The everyday connotations of 'believes' are relevant only to the extent to which they are consistent with this technical meaning.

Formally the belief relation of a language automaton is a two-place relation $B \subseteq Z \times S$ defined by the condition that $B(z, s)$ if and only if $\lambda(z, \langle s, ?\rangle) = \mathbf{Y}$. '$B(z, s)$' may be read "A language user in state z would believe s" or more briefly "z believes s".

EXAMPLE 2.2. *Learning and Belief in the Detectives' Language*. The learning operation and belief relation of the language automaton of Figure 2.2 are displayed in tabular form in Table 2.1. The derivation of the table from the diagram should be obvious. A blank entry indicates a state-sentence pair for which L or B has been left undefined. This tabular representation is an adaptation of the automata theorist's 'flow tables', and is an alternative way of specifying a language automaton.

A language automaton is uniquely characterized by its sentence set S, its state set Z, its learning operation L, and its belief relation B. It can therefore be represented as a 4-tuple $\langle S, Z, L, B\rangle$ as well as a 5-tuple $\langle X, Y, Z, \delta, \lambda\rangle$. The 4-tuple representation departs from standard conventions of automata theory but is so much simpler that it will be adopted in most of the sequel.

Some terminological remarks: The word 'accepts' would be suitable as an informal reading for L or B were it not ambiguous between the two. In its dynamic sense in which it connotes the act of receiving and assimilating new information, 'accepts' is a good reading for L; thus '$L(z, s)$' could be read "the state entered upon accepting s while in z". But 'accepts' also has a static sense, e.g. '$B(z, s)$' may be read "language users in state z (already) accept what s says". Because this ambiguity blurs the very distinction between B and L which should be emphasized, the 'learns' and 'believes' readings seem preferable, though unfortunately 'believes' suffers to some extent from the same ambiguity.

In R. M. Martin's system of pragmatics (1959) 'accepts' is taken as the principal pragmatic concept. From his discussion it is clear that Martin intends

TABLE 2.1

	L				*B*			
	Coe was in	*Coe was out*	*Doe was in*	*Doe was out*	*Coe was in*	*Coe was out*	*Doe was in*	*Doe was out*
z_1	z_4	z_2	z_3	z_5	N	N	N	N
z_2	—	z_2	z_6	z_9	N	Y	N	N
z_3	z_7	z_6	z_3	—	N	N	Y	N
z_4	z_4	—	z_7	z_8	Y	N	N	N
z_5	z_8	z_9	—	z_5	N	N	N	Y
z_6	—	z_6	z_6	—	N	Y	Y	N
z_7	z_7	—	z_7	—	Y	N	Y	N
z_8	z_8	—	—	z_8	Y	N	N	Y
z_9	—	z_9	—	z_9	N	Y	N	Y
z_{10}	z_7	z_2	z_3	z_9	N	N	N	N
z_{11}	z_4	z_6	z_3	z_8	N	N	N	N
z_{12}	z_4	z_9	z_7	z_5	N	N	N	N
z_{13}	z_8	z_2	z_6	z_5	N	N	N	N
z_{14}	z_7	z_9	z_7	z_9	N	N	N	N
z_{15}	z_8	z_6	z_6	z_8	N	N	N	N

'accepts' to be interpreted in a sense similar to our 'believes' (pp. 10ff. and 33ff.). However, in Martin's system 'accepts' is a three-place relation read

P accepts *s* at time *t*

where *P* is a person, *s* a sentence, and *t* a 'time-slab'. Our belief relation is a two-place relation readable as

People in state *z* believe *s*.

For present purposes at any rate, the two-place relation is more convenient because it does away with the bothersome necessity of relativizing all pragmatic statements to particular language users at particular times.

When the intended physical interpretation of a language automaton happens to be a computer rather than a human, all of the terms 'learns', 'believes', or 'accepts' seem objectionable to the extent that they anthropomorphize a machine. There may even be related philosophical objections when the intended interpretation is the linguistic competence of a human. In any case the English readings must be regarded as having at best an approximate mnemonic worth.

2.5. Black-box Methodology

Some methodological issues concerning the nature of a 'black-box' analysis need to be discussed at this point. When an organism or other physical system is to be described, its internal mechanisms are often unobservable, difficult to observe, or of no interest. In such circumstances the system may be viewed as a 'black box' whose real internal structure is to be ignored. Many psychological investigations are black-box analyses and one suspects investigations in other sciences are of this character more often than is recognized.

There are two senses in which an automaton can serve as a mathematical model of a physical system. In a true 'black-box' analysis the mathematical automaton is intended only as a *behavioral model* of the system. In other ('glass-box') investigations it may be intended as a *structural model* too, meaning that the internal structure of the system as well as its behavior is supposed to be reflected in the mathematically defined automaton. When an automaton is intended only as a behavioral model, its internal details are hypothetical and do not matter so long as they give rise to the right input and output. When it is intended as a structural model, its structure does matter because it is intended to tell something about the system's physical structure.

In a black-box analysis – an analysis whose goal is to produce a behavioral model – there is no uniquely correct model. For every automaton there exist other automata differing from it in internal structure but identical to it in behavior. Thus when someone describes an automaton as a behavior model of an observed system, it is always possible to describe other automata behaviorally indistinguishable from it and hence equally valid as behavioral models. In a black-box analysis the choice between indistinguishable models is arbitrary and is usually based on secondary criteria such as simplicity or adherence to historic convention.

Clearly then, behavioral indistinguishability is a crucial notion underlying all black-box analyses. Fortunately it is well understood in automata theory. Two automata that are indistinguishable in their input–output behavior are said in automata theory to be 'equivalent'. In this terminology the aim of a black-box analysis is to specify a model which is descriptively accurate 'up to equivalence', or in more general terms to specify the equivalence class to which the system of interest belongs. The equivalence concept is absolutely basic, for if it is not clearly understood that an analysis is supposed to have validity only up to equivalence, vacuous disputes can arise about which of several behaviorally indistinguishable analyses is 'correct'.

Equivalence among automata is defined in terms of equivalence among states, for which some special notation is needed. The state transition function δ and output function λ of an automaton can be extended to functions $\bar{\delta}$ and $\bar{\lambda}$ which apply to finite sequences of inputs rather than single inputs in obvious ways. The extended transition function $\bar{\delta}$ then specifies, for any finite non-null sequence $\bar{x}$ of inputs, the new state $\bar{\delta}(z, \bar{x})$ entered if the automaton initially in z receives the inputs one at a time in the order of $\bar{x}$. Similarly the extended output function $\bar{\lambda}$ specifies, for any z and finite non-null input sequence $\bar{x}$, the final output $\bar{\lambda}(z, \bar{x})$ obtained when the automaton, initially in state z, is fed $\bar{x}$. (Formally $\bar{\delta}$ and $\bar{\lambda}$ are defined inductively by $\bar{\delta}(z, \bar{x}) = \delta(\bar{\delta}(z, \bar{x}'), x)$ and $\bar{\lambda}(z, \bar{x}) = \lambda(\bar{\delta}(z, \bar{x}'), x)$ for all $\bar{x}$ consisting of $\bar{x}'$ followed by x (Hartman and Stearns 1966, pp. 22, 23).) Equivalence among states is now definable as follows.

DEFINITION 2.4. Let $\mathcal{A} = \langle X, Y, Z, \delta, \lambda\rangle$ and $\mathcal{A}' = \langle X, Y', Z', \delta', \lambda'\rangle$ be automata with the same input set X. Then for any $z \in Z$ and $z' \in Z'$, z is *state-equivalent* to z' (symbolically $z \simeq z'$) if and only if for every non-null sequence $\bar{x}$ of members of X,

$$\bar{\lambda}(z, \bar{x}) = \bar{\lambda}'(z', \bar{x}).$$

The definition makes states state-equivalent just in case their automata, if initially set in those states, would be behaviorally indistinguishable. (In incomplete automata the last line of the definition is interpreted to mean that $\bar{\lambda}(z, \bar{x})$ and $\bar{\lambda}'(z', \bar{x})$ are either both undefined or else both defined and equal.)

Where the automata in question are language automata, an extended learning operation $\bar{L}$ and extended belief relation $\bar{B}$ are definable in a manner analogous to the way in which $\bar{\delta}$ and $\bar{\lambda}$ were defined for arbitrary automata. For two language automata $\langle S, Z, L, B\rangle$ and $\langle S, Z', L', B'\rangle$ the definition then asserts that $z \simeq z'$ if and only if for every finite non-null sequence $\bar{s}$ of sentences from S,

$$\bar{B}(z, \bar{s}) \text{ if and only if } \bar{B}'(z', \bar{s}).$$

Intuitively, two information states are state-equivalent if and only if it does not matter which state one starts out in; learning the same series of sentences always puts one in the same final state.

Whole automata can now be defined to be equivalent just in case they could always be made to mimic each other by appropriate choice of initial state.

DEFINITION 2.5. An automaton $\mathfrak{A} = \langle X, Y, Z, \delta, \lambda \rangle$ is *equivalent* to another automaton $\mathfrak{A}' = \langle X', Y', Z', \delta', \lambda' \rangle$ (in symbols $\mathfrak{A} \cong \mathfrak{A}'$) if and only if

(i) $X = X'$ and $Y = Y'$;

(ii) for every $z \in Z$ there exists a state-equivalent $z' \in Z'$, and for every $z' \in Z'$ there exists a state-equivalent $z \in Z$.

In case both automata happen to be language automata, (i) can be shortened to read simply $S = S'$.

An automaton is *reduced* if and only if no two of its distinct states are state-equivalent. In the case of finite automata it is well known that for every automaton there exists an equivalent reduced automaton that is unique up to isomorphism. It might have been hoped that by confining all behavioral models to reduced automata one could escape or ameliorate the problem of non-uniqueness of behavioral models. But unfortunately even isomorphic automata can differ radically in the way their internal states are represented, and reduced automata do not always provide the most enlightening models anyway. It would therefore seem that the existence of many equivalent behavioral models for any given system is simply an unavoidable complicating factor inherent in the black-box approach.

2.6 THE WHAT-DO-YOU-KNOW? GAME

The concept of a behavioral model allows a further clarification of the notion of a language automaton for a particular language. Let us consider an imaginary parlor game called 'What-Do-You-Know?' In the game there are two contestants called the Questioner and the Answerer. A game consists of a dialogue between the Questioner and the Answerer in which the Questioner is allowed to make any statements he pleases and to ask questions starting out with the phrase 'DO YOU KNOW THAT . . .'. The Answerer does nothing but give yes-or-no replies to the Questioner's questions. The idea of the game is that the Answerer must try to give 'correct' answers based on his own knowledge amended by whatever the Questioner has told him so far. To do so the Answerer must accept all the Questioner's statements, or at least pretend to do so for the sake of the game.

Here is a sample game.

Q: *All birds are bipeds.*
Q: DO YOU KNOW THAT *birds are all bipeds?*
A: Yes.

Q: DO YOU KNOW THAT *there are no birds that are not bidpeds?*
A: Yes.
Q: DO YOU KNOW THAT *diurnal birds of prey are bipeds?*
A: Yes.
Q: DO YOU KNOW THAT *birds are carnivorous bipeds?*
A: No.
Q: *Bipeds are vertebrates.*
Q: DO YOU KNOW THAT *every bird is a vertebrate?*
A: Yes.
Q: *Humans are mammals.*
Q: *Mammals are vertebrates.*
Q: DO YOU KNOW THAT *birds are mammals?*
A: No.

In this particular game it would probably be agreed that the Answerer has answered all his questions correctly and so deserves a perfect score. The object of the game from the Questioner's point of view need not concern us.

During the course of a What-Do-You-Know? game the Answerer acts the part of a language automaton. In fact, he *is* a realization of a language automaton so long as he plays correctly. The Questioner can, as it were, store information in him and later deduce the properties of the resulting information state by submitting statements prefixed with DO YOU KNOW THAT to him as test inputs.

EXAMPLE 2.3. *What-Do-You-Know? in the Detectives' Language.* Suppose a detective currently in state z_{10} of the language automaton for the detectives' language (Figure 2.2, p. 23) were asked to be the Answerer in a game of What-Do-You-Know?. Here is a game that might result.

Q: DO YOU KNOW THAT *Doe was in?*
A: No.
Q: DO YOU KNOW THAT *Doe was out?*
A: No.
Q: *Coe was in.*
Q: DO YOU KNOW THAT *Doe was in?*
A: Yes.

Now let us imagine what could never really exist, a perfect What-Do-You-Know? player capable of playing the role of Answerer flawlessly no matter what the Questioner asked. No matter how long or complex the Questioner's statements, this Answerer never makes a slip so long as the questioning is carried out in good English. He does not have to be knowledgeable at the start of his games, but when told something during a game he assimilates it and makes full use of it henceforth. His answers do not have to be right in the absolute sense (the Questioner is allowed to give him false information), but must be correct relative to what he has been told. If asked questions the answers to which are not determined by what has gone before in the game, his answers are determined by what his initial beliefs were at the start of the dialogue.

The notion of an ideal Answerer embodies the idea of a language automaton for a language. *A language automaton for a particular language is a behavioral model of an ideal What-Do-You-Know Player in that language.* The input–output behavior of a language automaton for a language is indistinguishable from that of a user of the language when playing What-Do-You-Know? correctly, and this may be taken as a working *definition* of what a language automaton for a language is. It implies that for any initial information a hypothetical perfect player might possess, the automaton could be set in an equivalent starting state such that player's and automaton's answers would be identical for any line of questioning that might ensue; and conversely. This is of course just the definition of a behavioral model. And just as there is only one correct way of playing What-Do-You-Know? in a given language, so the language automaton for a given language is unique up to equivalence.

Someone might object: "A What-Do-You-Know? player obviously has to do some deductive logic, but nothing was said about a language automaton's having deductive components. Where did the logic creep in?" The logic did not creep in any back door but was implicit from the start in the concept of an information state. The science of information was not claimed to be independent of the study of logic; in fact information and logic are inextricably linked in a way which will become clearer in later chapters.

Another objection might be that a deterministic automaton which acts like a perfect What-Do-You-Know? player is impossible in principle for sufficiently rich languages according to classical undecidability results. This objection would be valid if the definition of a language automaton had been restricted to finite automata or Turing machines. But undecidability considerations do not prevent the representation of a perfect player as an arbitrary abstract automaton.

2.7. THE BEHAVIOR-ANALYTIC INTERPRETATION OF LANGUAGE AUTOMATA

Why should a behavioral model of an idealized player of an imaginary parlor game be taken seriously as a focal point for language analysis? It will be argued next that although a language automaton may not model any important language-associated system in its entirety, it nevertheless models a linguistic *component* of no less a system than the language user himself. The argument is that when one attempts a behavioral analysis of a language user with emphasis on linguistic behavior and with appropriate idealizations, one is led eventually to study a behavioral substystem of the user with the processing capabilities of a language automaton.

Suppose a linguistic community is under observation and it is desired to make a behavioral analysis of a typical member of the community with emphasis on his verbal behavior. The user may be regarded in the most general terms as a system with sensory input and motor output as shown in Figure 2.3. The sensory input includes not only linguistic input but all other sights, sounds, etc. that affect the user's behavior, and the motor output includes all his actions. To anyone but the most ardent behaviorist, such a diagram is apt to seem vacuous if not dehumanizing, but it is the unavoidable starting point of any behavioral analysis.

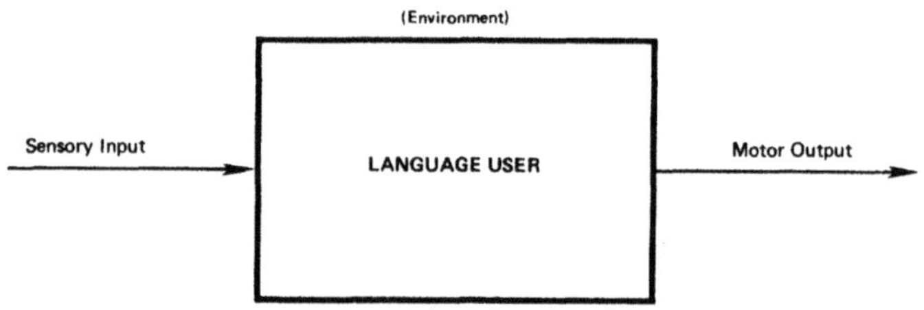

Figure 2.3. General behavioral view of the user of a language.

So far as the language scientist is concerned, the language user is a black box or something close to it. The human system as a whole would not ordinarily be studied as though it were a black box since much is known about structural anatomy, but in a study of sophisticated verbal behavior the neurophysiological facts presently available offer so little assistance to the investigator that he must rely mainly on the black-box approach. Questions of speech-sound production are of course physiological matters concerning which no one would think of conducting a black-box analysis. But on the level of syntax and beyond, physiology has had few insights to offer. Moreover, even if all the relevant neurological details were available to explain the advanced linguistic functions, the black-box approach would still be a legitimate one where the goal is simply to arrive at a codification of the behavioral conventions of the language. Just as the rules of arithmetic can be stated without a detailed account of how an electronic calculator operates, so it is legitimate to attempt to state rules of linguistic communication without describing their neurological implementation.

An obvious first move for the behavioral analysis is to distinguish between linguistic and nonlinguistic input and output. Some of the language user's inputs and outputs are linguistic utterances and some are not, and no investigation of verbal behavior can get off the ground without the assumption that the two can be distinguished by workable criteria of some sort. Let us assume then that linguistic inputs and outputs are distinguishable from nonlinguistic and that they have the form of sentences. Next, an analogous assumption is needed for the interior of the black box. It must be assumed that the rules of verbal behavior are somehow distinguishable from the other behavioral rules governing the organism, for if this assumption is not allowed it would be meaningless to attempt to focus attention on verbal behavior alone. The working assumption to which one is led is that a language user's behavior is separable into verbal and nonverbal subsystems, or at least that no conflict with observation will arise if the interior of the black box is hypothetically subdivided in this way.

Figure 2.4 shows the result. The black box is shown with input–output of two distinct types and with a corresponding interior partition. There is of course interaction across the partition indicated by the vertical arrows. It is not clear yet what this interaction is like and there may be room for considerable latitude in the form these signals are assumed to have. For that matter, there may be room for a great deal of arbitrary choice in the task assignments of the Verbal and Nonverbal Subsystems. Also, the fact that in the diagram the verbal input appears to go only into the Verbal Subsystem and the nonverbal into the Nonverbal Subsystem should not be taken too seriously, since either kind of input could reach either subsystem *via* the vertical arrows. All in all, the assumptions implicit in Figure 2.4 are really rather frugal.

The Verbal Subsystem must now be broken down into separate tasks in some helpful fashion. One might start at the left end of the box by looking into the effects produced by an incoming sentence. It has to be recognized that not all sentences produce the intended effect on their receiver, and some have no significant effect. The simplest reason for the failure of an utterance

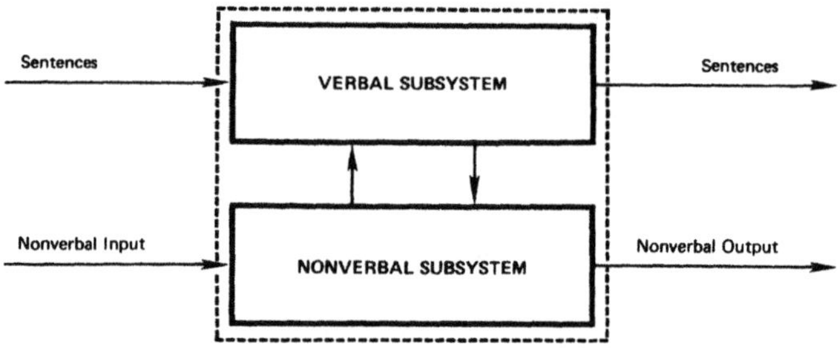

Figure 2.4. Black-box diagram of the language user with verbal and nonverbal behavioral rules assumed to be separable.

to have effect is that the utterance was simply not believed. At some point in the communication process a decision is evidently made for each sentence received about whether what the sentence says should be accepted or not. If we assume that this point comes sooner rather than later, as seems natural, we are led to postulate that upon entering the Verbal Subsystem every sentence is subjected to acceptability tests of some sort.

The acceptability tests may be called 'input selection criteria' since their task is to select the sentences worthy of being taken seriously and screen out the rest. It is convenient to assume that the input selection is carried out by an isolable subcomponent of the Verbal Subsystem called the 'Input Selector', and that incoming sentences reach the Input Selector before they get to any other part of the Verbal Subsystem. This assumption is somewhat arbitrary, but in a black-box analysis the sequence of internal operations is arbitrary so long as the results agree with observation.

The human input selector doubtless employs acceptance criteria of great delicacy and sophistication, basing its decisions not only on the sentence itself but also on data received from elsewhere, including the Nonverbal Subsystem. The form of its output is simple, though, and can be thought of as either the input sentence or the null output depending on whether the sentence satisfies the selection criteria or not. The possibility of a continuum of degrees of believability will be discussed later, as will certain interpretive tasks best assigned to the Input Selector.

There is a somewhat parallel situation with regard to outgoing sentences. At any given moment the language user has the choice of uttering any sentence or remaining silent. This choice is made on the basis of 'output selection' criteria whose task is to select from among all the possible sentences of the language the one to be uttered, if any. In human verbal behavior the output selection criteria involve not only myriad social conventions about what to say and when, but also higher goal-direction intellectual activity bearing on what utterances it would be most profitable to make to obtain currently desired ends under currently existing conditions.

But of all the factors bearing on what a language user elects to say, there is one of special concern here. In deciding whether to utter a sentence, the speaker must surely take into account whether he believes what the sentence says. Unless he is an inveterate liar or bluffer, a sentence he believes is a better candidate for utterance than one he doesn't believe, other things being equal. Hence the output selection criteria must make use of data about which sentences are believed on the basis of the language user's current information.

For the sake of the black-box model, the output selection criteria may be imagined to be assembled together into an 'Output Selector' serving as the final output stage of the Verbal Subsystem. The input to the Output Selector may come from any or all other components, but for definiteness the indications of which sentences are currently believed may be conceived as reaching the Output Selector from some other component in response to sentences submitted to it for this purpose by the Output Selector. Other conventions could of course be adopted without changing anything essential.

The result of postulating the existence of an Input and an Output Selector is shown in Figure 2.5. There the Verbal Subsystem is broken down into three subcomponents with the Input Selector serving as an input screening device, the Output Selector choosing the uttered output, and the residue brought together under the label 'Central Verbal Subsystem'. Interaction could exist between any two of the total of four entities so produced, but because the nature of most such interactions is unclear they are represented in the diagram by unlabelled arrows. The labelled arrows are the interactions of interest already mentioned: specifically, the labelled arrow from the Input Selector to the Central Verbal Subsystem represents accepted sentences and the labelled arrows between the Central Verbal Subsystem and the Output Selector represent candidate sentences submitted elsewhere by the Output Selector to see if they are currently believed and the believed/nonbelieved responses received in response.

Since Figure 2.5 is a black-box analysis, nothing is implied about its uniqueness or its structural correspondence to what really goes on inside a language user's head. Certainly the analysis is not claimed to be an *efficient* possibility; to the contrary, the assumed interaction between Output Selector and Central Verbal Subsystem (for example) is so obviously inefficient that it could not have much psychological or physiological reality. But as one way of

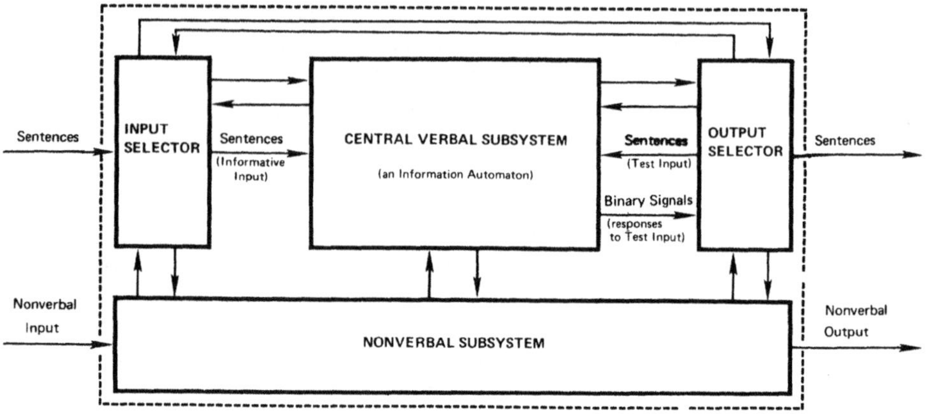

Figure 2.5. Same black-box diagram with rules of input and output selection assumed separable from the remainder of the Verbal Subsystem.

compartmentalizing tasks which is at least possible in principle, the breakdown has plausibility. Something corresponding to input and output selection must surely take place at some point in speech behavior, so the only new assumption is that the selection tasks can be consolidated, for the convenience of a black-box schematization, at the input and output ends of the Verbal Subsystem. The forms of internal interaction shown as labelled arrows are arbitrary conventions and a number of alternative possibilities could have been substituted without altering the essential content of the scheme.

We will not be concerned with the Input or Output Selector *per se*, but for the sake of a clearer understanding of what is being excluded from consideration it may be helpful to explore in more detail what the Selectors have to do. The Input Selector, which must decide which incoming sentences are to be learned, has to be equipped with the means of determining first just what information the sentences were intended to convey. For ambiguous sentences this implies a capability for breaking ambiguities. The Input Selector must therefore be assumed to contain subsystems capable of 'disambiguation' and possibly of other interpretive tasks as well. Next, and as its principal task, the Input Selector must apply to the new disambiguated sentences criteria for deciding whether the sentence should be accepted or not, or in a more general formulation to be taken up later the degree to which they should be accepted. The criteria must be sophisticated enough to take account of the probable goals and intentions of the utterer of the sentence in order to assess the credibility of the sentence. MacKay (1969, p. 127) has argued that in order to understand speech listeners must have "the ability to embody a skeleton representation of the goal-directedness of their interlocutor and his aims in using words". If MacKay is right, nothing less than this order of sophistication is involved in the Input Selector.

Similar remarks apply to the Output Selector. If what a speaker says is going to be of real service to his listener, he will have to make sophisticated inferences about his listener's goals and needs. Hence the Output Selector too must have access to "a skeleton representation of the goal-directedness of the interlocutor". But the situation is even more complicated in the case of the Output Selector, for in deciding what to say a speaker must take his own goals into account too. We conclude that both Selectors are incredibly complex and that their analysis might well be put off until some of the more straightforward aspects of linguistic communication have been explored more thoroughly.

With the Input and Output Selectors separated off, the big question concerns the nature of the Central Verbal Subsystem. How much or how little work it is given depends of course on how the line is drawn between Verbal and Nonverbal Subsystems. An attractive way of assigning tasks consistent with everything assumed so far is to postulate that *the Central Verbal Subsystem is an information automaton*. Under this assumption the current state of the language user's knowledge is represented in the state of this central component; his information is augmented and changed by informative inputs delivered to this component; and his current information affects his speech and actions by virtue of test inputs sent to it and its output responses.

Some of the Central Verbal Subsystem's input–output is linguistic; it takes the form either of sentences or of responses to sentences. The rest (in the diagram, the unlabelled arrows) may be classed as nonlinguistic and consists of sensory stimuli or processed remnants thereof and signals for candidate motor signals other than speech signals. The nonlinguistic input–output has an unknown and presumably problematic form. Now, if attention is restricted to the linguistic input–output, the result is a language automaton. Thus *a language automaton for a language is obtainable as a subautomaton of the information automaton serving as Central Verbal Subsystem, namely the subautomaton obtained by restricting to linguistic input–output*. Every language user carries a language automaton around in his head, or at least his verbal behavior suitably idealized can be analyzed as though he did.

To help clarify the relationship of this language automaton to the information automaton in which it is embedded we state here the standard definition of a 'subautomaton':

DEFINITION 2.6. An automaton $\langle X', Y', Z', \delta', \lambda' \rangle$ is a *subautomaton* of another automaton $\langle X, Y, Z, \delta, \lambda \rangle$ if and only if

(i) $X' \subseteq X$;

(ii) $Y' \subseteq Y$;

(iii) $Z' \subseteq Z$;

(iv) δ' is the restriction of δ to $X' \times Z'$;

(v) λ' is the restriction of λ to $X' \times Z'$.

A subautomaton $\mathfrak{A}' = \langle X', Y', Z', \delta', \lambda' \rangle$ of $\mathfrak{A} = \langle X, Y, Z, \delta, \lambda \rangle$ is called an *X-subautomaton* of $\mathfrak{A}$ if and only if $Y' = Y$ and $Z' = Z$; a *Y-subautomaton* of $\mathfrak{A}$ if and only if $X' = X$ and $Z' = Z$; a *Z-subautomaton* of $\mathfrak{A}$ if and only if $X' = X$ and $Y' = Y$; an *X,Y-subautomaton* of $\mathfrak{A}$ if and only if $Z' = Z$; and so on (Starke 1969). In this terminology the language automaton of interest is an X,Y-subautomaton of the Central Verbal Subsystem.

This statement has to be qualified a little insofar as the Central Verbal Subsystem cannot be said for sure to be wholly sequential and deterministic; it is possible that some continuous nonlinguistic input should be allowed for and possibly some indeterminacy. The language automaton embedded in it would remain discrete and determinate though.

The language automaton obtained directly as an X,Y-subautomaton of the Central Verbal Subsystem is unnecessarily complex, containing as it does many states that were inequivalent when there was nonlinguistic input–output but which are now state-equivalent. The simplest and most enlightening kind of language automaton for a language is for this reason probably not the X,Y-subautomaton itself but a more streamlined language automaton equivalent to it. All that is needed is a behavioral model of this X,Y-subautomaton so it need be described only up to equivalence.

This analysis makes it clear how state changes can occur in the Language Automaton within the Verbal Subsystem without any linguistic messages being received as input. Such changes can come about in response to nonlinguistic input to the Central Verbal Subsystem of which the Language Automaton is a subautomaton. But changes of this kind are of no interest when it is the Language Automaton itself that is the object of study. A language user's internalized language automaton changes its information state frequently, perhaps constantly, in response to nonlinguistic as well as linguistic input, but only the changes caused by linguistic input are studied by the language scientist on the level of analysis that is of concern here.

2.8 THE LINGUISTIC PRIORITY OF THE LANGUAGE AUTOMATON

The behavioral analysis of the preceding section showed the Language Automaton to be an isolable subsystem of verbal behavior, but along the way it also turned up other isolable subsystems such as the Input and Output Selectors and the Central Verbal Subsystem in which the Language Automaton is embedded. In what sense is the Language Automaton more important linguistically than these?

Let us speak of a rule governing a given language user's verbal behavior as a *universal* of his language provided it also governs every other user of the language. Such a property might conveniently be called an *intra-language* universal to distinguish it from the 'inter-language' universals more often discussed by linguists, an inter-language universal being a property shared by all natural languages. Now, it seems natural to suppose that *the intra-language universals of verbal behavior (some of which may also be inter-language universal) constitute the principal subject matter of language science*. This principle seems to be embraced by almost all schools of linguistic thought, for it is implicit in the very concept of a language as a system of shared rules of communication.

To elaborate, what must be described in connection with a language is sometimes said to be a 'linguistic competence' common to all users of the language. In the presence of appropriate idealizations, this competence can be viewed as a set of laws describing the verbal behavior of the language user whenever he uses his language properly, so that a linguistic competence is a set of intra-language universal characteristics of ideal verbal behavior. There may of course be minor discrepancies, due to differences of dialect and idiolect, between the rules followed by one language user and those followed by another, but the ideal situation toward which all the language users strive is the sharing of a common linguistic competence.

There is nothing surprising in the linguist's preoccupation with intra-language universals. From Aristotle through Wittgenstein, philosophers of language have stressed the role of shared conventions as the basis of linguistic communication – of the necessity for all players of the language game to play by the same, or approximately the same, rules. The controversy is about the *nature* of the shared conventions; there is little doubt that whatever they are they must be essentially universal among all speakers of the same language. The language can in fact be regarded as just the sum of all such common conventions.

Now, the language automaton of one user of a language is behaviorally the same as the language automaton of any other user of the same language: people who speak the same language have equivalent language automata. This equivalence is evident from the fact that two different What-Do-You-Know? players, both ideal, who speak the same dialect will always give the same answers when playing the game correctly starting with the same initial information. This observation shows the intra-language universality of that part of a language user's linguistic competence representable as his language automaton. It also shows the language automaton to be a legitimate part of the principal subject matter of language science.

It is also noteworthy that, of the various subsystems postulated by the behavioral analysis, *only* the Language Automaton is wholly intra-language universal. The Input and Output Selectors and Central Verbal Subsystem may have parts that are universal, but none of them is entirely so. Consider the Input Selector, for instance. The criteria governing whether a received linguistic message is believed or not seem to vary greatly from individual to individual: one person may use mainly intellectual criteria in deciding whether to believe an assertion, while another looks only at the mendacious expression on the speaker's face; one may be gullible while another is skeptical; and so on. Such considerations suggest that there is nothing like complete uniformity among the input selectors of different language users; at the very least, it is far from obvious how all input selectors could be meaningfully regarded as embodying exactly the same acceptance criteria. It was obvious from the nature of the What-Do-You-Know? game that all speakers of the same language must have essentially equivalent language automata, but there is no corresponding argument for input selectors. To the contrary, the longer one examines input selection criteria the more non-universals one finds.

This is even more true of output selectors. Whether a speaker elects to say something at a given moment and what he elects to say are governed by his mood, physical well-being, intellectual interests of the moment, his native loquaciousness or lack of it, his current strategies for getting what he wants, plus countless other personal factors which are not universal but highly idiosyncratic. There may be some strained sense in which different speakers' output selectors could be considered equivalent in principle with all observed differences due merely to their different internal states, but this would make the state sets very exotic indeed. So far as any workable analysis is concerned, different language users have very different output selection criteria.

That a speech community's language automata should be equivalent but their selectors idiosyncratic is not really very surprising. How the sentences of a language convey information is a matter of arbitrary convention and everyone had better abide by the same conventions or there will be no communication. But for deciding what heard sentences to accept and what believed sentences to utter, the language itself offers no inviolable rules; successful information transfer is in no way dependent on all users applying identical criteria for what to accept or transmit.

Taken as a whole the central verbal subsystem also seems non-universal. Far from being equivalent, it is not even clear that the central verbal subsystems for different speakers share the same input sets, the nonlinquistic input codes being possibly idiosyncratic. It would appear therefore that *of the subsystems considered, only the Language Automaton is essentially equivalent for different speakers of the same language* in a meaningful sense. Hence only the language automaton is entirely concerned with the principal subject matter of linguistics.

Of course, the fact that the other components are not entirely universal does not rule out the possibility that they might be partly so. It is also possible that some radically different black-box analysis might produce a different partitioning which was also consistent with observation but had other universal components. The present analysis is not claimed to be the only one possible; all that is suggested is that it yields a self-consistent breakdown and that the Language Automaton is the only wholly intra-language universal element to emerge from it.

That other components might be *partly* universal is not just a theoretical possibility; it is demonstrable that every language *must* have at least some universals beyond those embodied in its language automata, or meaningful communication could not take place. To see this, compare English with a variant language obtained from English by switching the meanings of some grammatically similar words, say by permuting the color terms or reversing *in* and *out*. English and the variant language would presumably have equivalent language automata, but are obviously different languages in the common sense meaning of 'language': speakers of the one would have trouble communicating with speakers of the other on the subject of colors or who was in and who was out. It follows that at least some universal conventions must be lost in going from a language to those aspects of it captured in its language automata.

It is even possible to point to where the main loss occurs. It happens when attention is restricted from the entire Central Verbal Subsystem to the Language Automaton embedded in it. When the nonlinguistic input–output of the larger information automaton is done away with, all contact with the physical world of (nonliguistic) sense data is lost. The Language Automaton that remains represents the 'inner structure' of the language but tells nothing about the relationship of the language to the outside world.

This restriction of attention to inner structure is not necessarily bad; in fact, it is a resolution of Bloomfield's Dilemma. That Dilemma, it will be recalled, was the problem of how to delimit the bounds of a theory of language in a natural way without confining the theory to the syntactic level. Restricting attention to the Language Automaton breaks the Dilemma by isolating an aspect of linguistic behavior that goes beyond syntax without committing the investigator to an analysis of all the knowledge brought to bear in using the language.

The critical step in isolating the Language Automaton, apart from the exclusion of the Selectors, was the focusing of attention on the Central Verbal Subsystem's linguistic input–output. The only methodologically delicate aspect of this step is the assumption that it is possible to distinguish the linguistic input–output from the nonlinguistic. But this distinction is inherent in all language investigations including purely syntactic ones. And since making the distinction has not proved an insurmountable obstacle to syntax, it is hard to see why it should present serious new difficulties when used to demarcate the bounds of a certain level of pragmatic analysis. Our resolution of Bloomfield's Dilemma therefore comes to this: We have taken the customary distinction between linguistic utterances and other kinds of sense data and motor actions and applied it a little deeper inside the black box.

2.9 LANGUAGES

We have seen that a language automaton embodies a coherent set of rules of verbal behavior, and that this set of rules is an appropriate first focus of interest for linguistic analysis beyond the syntactic level. This suggests that language automata might be exploited, not just as encapsulations of certain isolated manifestations of language usage, but as surrogates (on a certain level of analysis) for the language itself.

The first idea that comes to mind is to take a language automaton for a language as the representation of the language, i.e. to say that language automata are languages viewed abstractly. But this won't do because the representation would not be unique; for any given language there are many language automata. Moreover, to take an individual language automaton as a language representation would give undue prominence to the automaton's internal structure – structure which in a black-box investigation is only an arbitrary hypothetical construct. What is wanted is a representation that abstracts from all details of internal construction.

Now a language has many language automata but they are all equivalent. An obvious technical device for de-emphasizing the internal structure of individual language automata, therefore, is to associated with each language the set of *all* its language automata. This amounts to taking entire equivalence classes of language automata as language surrogates.

The equivalence-class mode of representation does indeed abstract from all internal details of automaton structure; it is in fact 'abstraction' in the classic sense of moving from a space to its partitioning ('quotient space') defined with respect to the equivalence relation. After the universe of language automata has been partitioned in this way, the resulting equivalence classes make up the linguistic universe of all logically possible languages.

An *equivalence class* with respect to an equivalence relation is a set of members of the domain of the relation such that every pair of members in the set is equivalent and none are equivalent to elements outside the set. Taking the equivalence relation to be automata-theoretic equivalence (Definition 2.5ff.), the central definition of the theory may be stated thus:

DEFINITION 2.7. $\mathfrak{L}$ is a *language* if and only if $\mathfrak{L}$ is an (automata-theoretic) equivalence class of language automata.

There is historical precedent for representing systems abstractly as equivalence classes in, for example, the Frege–Russell representation of the cardinal numbers as classes of cardinally equivalent sets.

The *sentence set* of a language $\mathfrak{L}$ is the sentence set S common to all the language automata in $\mathfrak{L}$. Different languages can have the same sentence set, reflecting the fact that the representational system goes beyond syntax.

In defining a 'language' to be an equivalence class of language automata, we are in essence suggesting that *languages may be analyzed by analyzing the behavior of their associated language automata*. In other words, it is proposed that language automata are legitimate objects of linguistic study, though their legitimacy resides only in their derivative capacity as surrogates for the equivalence classes to which they belong. Since this notion of 'language' may at first seem a little esoteric and elusive, it may be helpful to make a few comments on it here in anticipation of later developments.

1. The use of the equivalence-class concept in the definition is merely a technical device. It formalizes the agreement to regard the internal structure of language automata as purely hypothetical. If someone were to take individual language automata as his object of study rather than equivalence classes thereof, there could be no objection so long as he kept firmly in mind that for a given language there may be many different language automata which are equally legitimate behavioral models of it.

2. In case even individual language automata seem too esoteric, it should be remembered that their automata-theoretic form is inessential. We have seen that the language rules embodied in a language automaton can be stipulated just as well e.g. by describing a sentence set, state set, and accompanying learning and belief relations.

3. It might be thought presumptuous to call an equivalence class of language automata a 'language' when language automata clearly fall short of reflecting all aspects of linguistic behavior. However, the intended spirit of the definition is that such a class is an adequate surrogate for a language *on a certain level of linguistic analysis*. It is not claimed to capture all the connotations of the word 'language' as used in daily discourse.

An analogy with the older style of linguistic analysis *qua* syntactic analysis may be helpful here. On the syntactic level of analysis a language is often represented simply as a set of sentence-representing strings (sometimes accompanied by deep structures, etc.), and no one objects to the name 'language' understood as a technical term for this string set. By the same token there should be no objection to calling the much more inclusive representation of a language studied in the present theory by the technical name 'language'.

4. That the representation does not reflect all conceivable aspects of lingustic behavior is not necessarily a defect. To the contrary, it offers hope that the strategy of isolating a subsystem for special study has been successfully applied. It will be seen later that the representation takes language analysis substantially beyond syntax, and even beyond what is currently regarded as semantics, without becoming impaled on the other horn of Bloomfield's dilemma – the attempt to analyze all linguistic phenomena at once.

5. It might plausibly be objected that there is little psychological reality to the division between the language automaton and the other components of

verbal behavior (cf. Figure 2.5). A more accurate statement, however, would be that it is not yet clear how much psychological reality the division has; there might, for example, be minor changes of input–output conventions which would make the entire arrangement more realistic. But however that may be, the issue of psychological realism has only a very indirect bearing on an investigation of linguistic competence. This is a principle already implicit in much current linguistic practice.

6. On the matter of realism, another point must be made which is quite crucial. Suppose a hypothetical mechanism for a language automaton for a natural language is proposed which is hopelessley unrealistic in any physical or psychological sense – one that is infinite, for example. Must it be rejected automatically on account of its lack of realism? No, because under the proposed definition of language it is the equivalence class as a whole that is of ultimate interest rather than any single member of it. A language automaton which is not natural could have equivalence classmates which are. Thus even when psychological realism is an ultimate goal (which it need not necessarily be), an unrealistic mechanism with the right behavioral properties could still be of legitimate interest so long as it is remembered that it is only an indicator of an equivalence class.

7. For a facet of linguistic behavior to be of truly deep fascination and first-order importance, more is required than that it be isolable. It must also reflect the operation of some unifying principle, a core idea of universally acknowledged interest. Although it is not apparent at this point, I shall try to show in later chapters that there is such a core concept in the case of language automata. The concept in question is logic. It will emerge that a language automaton embodies just those aspects of verbal behavior needed to reflect the 'logical' structure of the language, where 'logical' is to be understood in a certain rather broad sense encompassing both inductive and deductive reasoning and other closely related phenomena. It is in its role as a logical entity – as a link between linguistic information transfer and the concept of rational behavior – that the language automaton will be seen to be of deepest philosophical and scientific interest.

2.10 SUMMARY

The starting point of the theory of language and logic with which we will be concerned is that languages can be represented abstractly as equivalence classes of language automata. This representation is not claimed to provide a complete encapsulation of everything that might reasonably be called linguistic competence, but is thought to encompass more than most mathematical language models that have been proposed, including all of traditional syntax and (as will become clearer later) model-theoretic semantics. Other aspects of language for which the representation may provide an appropriate object of study, including meaning and 'logical structure', will be explored in the sequel.

The language representation that has been proposed is motivated by the idea of information transfer among language users as the basic function of language. The representation can in fact be associated with a hypothetical component of a behavioral model of a language user engaged in linguistic information transfer. This component (the 'Language Automaton') is essentially the aspect of linguistic competence exhibited when a language user plays the What-Do-You-Know? game. Without denying that it might be rewarding to try to construct realistic (e.g. error-prone) performance models of the component in question, we will be concerned here instead mainly with competence models – with the linguistic abilities of 'ideal' speaker-hearers, as they say. This means that we will be less concerned with the physical or psychological realism of the hypothetical mechanisms to be tried out for this component than might otherwise be the case. This, combined with the fact that it is not the particular mechanism that is of ultimate interest anyway but only its equivalence class as a whole, allows for great latitude in the kinds of mechanisms that might fruitfully be explored.

CHAPTER 3

ON DESCRIBING LANGUAGES

Since descriptive linguistics is essentially the activity of writing language descriptions, ways must now be found of formulating descriptions of languages represented abstractly as equivalence classes of language automata. One needs to know first what sorts of symbolisms might be needed for the specification of such equivalence classes and some of the strategies available for organizing the symbolic descriptions. Then, assuming one wants *correct* language descriptions, it is important to know how the symbolic descriptions relate to observable evidence, how to gather such evidence, and how to test out the accuracy of a description against that evidence.

3.1 DESCRIPTIVE STRATEGIES

There are a number of strategies which might be followed to produce a symbolic specification of an equivalence class of language automata, but the most obvious entails the description of some particular language automaton in the class. Let us speak of a language automaton which is a member of a language as a *structuralization* of that language. The most straightforward way of describing a language is to describe one of its structuralizations and to declare the language of interest to be the equivalence class of that structuralization. This method produces what we will call a *structural language description*. By writing a structural language description the linguist can have his cake and eat it too: he can deal with a definite structuralization, knowing the while that its internal structure has no ultimate reality except insofar as it determines an equivalence class.

The main body of a structural language description consists of the detailed description of the structuralization in question. In an *informal* description this description could consist of a state diagram, a flow table, a careful English description of the structuralization, or any of the other devices used in automata theory to describe an automaton informally. In a *formal* description the automaton is described rigorously in a mathematical notation. A structural language description always ends with a statement, implicit for informal descriptions but preferably explicit for formal, that the language in question is the equivalence class to which the just-described structuralization belongs.

EXAMPLE 3.1. *Informal Structural Description of the Detectives' Language.* The Detectives' Language is the equivalence class of the language automaton of Figure 2.2.

EXAMPLE 3.2. *Formal Structural Description of the Detectives' Language.* Let us call the Detectives' language '$\mathfrak{L}_0$'. Let '$\frown$' denote the concatenation operation on linguistic strings and let '$X + Y$', where X and Y are sets of finite strings, denote the set of all sequences $x \frown y$ obtainable by concatenating a member of X with a member of Y in that order. (The + operation is encountered in mathematical linguistics and has been called 'interconcatenation' (Cooper 1964).) A formal description of $\mathfrak{L}_0$ combining these with standard set-theoretic operations follows. The strategy is to specify a structuralization $\langle S, Z, L, B\rangle$ of $\mathfrak{L}_0$ in steps (1)–(4), and to declare $\mathfrak{L}_0$ its equivalence class in the final line (5).

(1) $N = \{\textit{Coe, Doe}\}$

$C = \{\textit{was}\}$

$A = \{\textit{in, out}\}$

$P = C + A$

$S = N + P$

(2) $M = \{m \mid m: N \to P\}$

$Z = \{z \mid z \subseteq M \text{ and } z \neq 0\}$

(3) For all $z \in Z$, $n \in N$, and $p \in P$,

$$L(z, n \frown p) = \begin{cases} z \cap \{m \in M \mid m(n) = p\} & \text{if this set non-empty;} \\ \text{undefined} & \text{otherwise.} \end{cases}$$

(4) For all $z \in Z$, $n \in N$, and $p \in P$,

$B(z, n \frown p)$ if and only if for every $m \in z$, $m(n) = p$.

(5) $\mathfrak{L}_0 = [\langle S, Z, L, B\rangle]_{\cong}$.

This completes the language description.

Step (1) of the description analyzes the language on the syntactic level. It specifies the sentence set S, and in the process also specifies other syntactic classes to which it will be convenient to refer later on in the description. It uses the procedure of defining a succession of five form-classes, each of which is specified either by listing its members explicitly or by indicating how it may be constructed out of other form-classes that have been previously defined. N is the class of nouns in the Detectives' language, C is its copula, A its complement-adjectivals. P, the class of predicate phrases, contains the strings *was in* and *was out*. Finally S is specified as the interconcatenation of N and P. Ordinary phrase structure grammars differ from such a series of set-theoretic definitions only in superficial respects.

Step (2) is the crucial one in which it is decided how the states of the language automaton are to be represented mathematically. The representation on which (2) is based makes use of the notion of 'states-of-affairs' (not to be confused with information states). The possible states-of-affairs of concern to the detectives are the four possible situations: Coe in & Doe in; Coe in but Doe out; Coe out but Doe in; and Coe out & Doe out. These possibilities are conveniently represented as the possible functions from the class of nouns N into the class of predicate phrases P, so the first line of (2) specifies M to be the set of all such functions. A typical member m of M would be the function which maps *Coe* onto *was in* and *Doe* onto *was out*.

The second line of step (2) defines the state set Z of the language automaton to be the set of all non-empty subsets of M. Thus a state z consists of a set of one or more ways of assigning members of P to members of N. Intuitively, an information state is determined by one or more possible states-of-affairs, namely, the states-of-affairs which a detective in that state believes might be the case.

Step (3) defines the learning operation L to be a rule which selects the new state to be entered by discarding appropriate functions from the old. The rule is that for a sentence $n \frown p$, the functions to be removed from the old state are those which map n onto something other than p. Intuitively, learning a sentence involves rejecting all possible states-of-affairs with which the statement would conflict. If learning the statement would require rejecting *all* possibilities, though, the learning cannot take place and L is left undefined.

Step (4) defines the belief relation by the requirement that a sentence $n \frown p$ is believed in a state if and only if all the functions in that state map n onto p. Conceptually, a sentence is believed just in case no state-of-affairs that would make the sentence false is held to be one which might actually be the case.

Step (5) merely specifies the language of interest to be the class of all language automata equivalent to the one whose components are the S, Z, L, and B just defined. The square brackets are an algebraic notation denoting the operation of taking the 'coset' or equivalence class of the entity within with respect to the subscripted equivalence relation – in this case automata-theoretic equivalence.

It is noteworthy that if the Detectives' Language were to be expanded by increasing the number of suspects, the only change that would have to be made in the formal description would be the lengthening of the list of proper names specified in the right side of the first line. This stands in marked contrast to the changes that would be needed in the state diagram for the automaton, whose size and complexity would increase combinatorially with the number of suspects. A cardinal advantage of such mathematically formulated descriptions, then, is that they are capable of describing language structure compactly and in a manner that (except for the vocabulary lists themselves) is independent of vocabulary size. This advantage alone makes them preferable to informal descriptive methods based on state graphs or flow tables for all purposes other than the casual presentation of trivial examples.

Structural language descriptions are not the only possible kind. A property of automata may be said to be a *behavioral property* if and only if any two equivalent automata must either both have the property or else both fail to have it. A behavioral property of language automata is a property of languages because it does not distinguish among structuralizations of the same language. Now, by specifying enough behavioral properties one can specify a language. That is, by listing behavioral properties until all equivalence classes of language automata are excluded but one, a language is described. This strategy produces a *behavioral language description*.

A disadvantage of behavioral language descriptions is that it is not always self-evident that there exists one and only one equivalence class satisfying the list of behavioral properties. An auxiliary proof is for this reason often required to establish the legitimacy of the description. Another drawback is that behavioral language descriptions can be rather long and unwieldy even for fairly trivial languages. Nevertheless they are certainly possible in principle and they serve to define the opposite endpoint of the descriptive spectrum from structural language descriptions.

Descriptive techniques lying between the endpoints of the spectrum will be said to yield *quasi-behavioral language descriptions*. Such methods are neither purely structural nor purely behavioral but contain elements of both approaches. Examples will be encountered later as illustrations of axiomatic descriptive techniques.

3.2 Descriptive Equivalence

A general methodological precept applicable to language descriptions of all kinds is the idea of *descriptive equivalence.* Two language descriptions are descriptively equivalent if and only if they describe the same language. Thus equivalent descriptions specify the same abstract object, and if the deductive apparatus of the metalanguage is rich enough, the one is mathematically derivable from the other. There can be no valid empirical grounds for preferring one language description over another one descriptively equivalent to it, since they assert the same thing about observable verbal behavior. However, there may be other grounds for preference such as simplicity, brevity, adherence to familiar descriptive conventions, or the capacity to give the reader clearer 'insight' into the character of the language.

Descriptive equivalence should not be confused with black-box (i.e. automata-theoretic) equivalence. Descriptive equivalence is a relation between symbolic, metalinguistic, specifications of equivalence classes of automata while black-box equivalence is a relation between the automata themselves. The two are however related in that structural descriptions specifying equivalent structuralizations are descriptively equivalent.

3.3 Language Descriptions as Scientific Theories

The nature of language descriptions becomes much clearer when they are viewed in the light of what is known about the scientific method in general. A description of a language amounts to a scientific theory about that language, so to understand it better it is appropriate to review the nature of all scientific theories. (The scientific method is of course relevant only to the description of languages whose rules must be deduced from observation; artificial languages have rules that are created by fiat and hence already known.)

The most widely accepted account of how scientific theories are verified or rejected is the so-called 'hypothetico-deductive method'. It is not the only account, but others tend to be either variations on the hypothetico-deductive theme or else are concerned with theory creation rather than theory justification. For the sake of having a definite theory of scientific methodology on the table as a basis for discussion, we summarize here the hypothetico-deductive method more or less as it was set forth in 1934 by Karl Popper (1959). It consists of the following four steps, though not necessarily in temporal order:

1. *The scientist formulates a theory*. The theory is a statement or set of statements about something of interest to the scientist. It is usually called a 'hypothesis' or 'conjecture' but occasionally a system of scientific 'postulates'. The theory must be internally consistent, but not tautologous. It has normally the character of a general or universal statement. A good theory is precise; ideally it is formulated or could be readily reformulated in one of the exact formal languages of mathematics or logic.

2. *The scientist deduces observational consequences from the theory*. The derived statement, or 'predictions', follow from the theory by the laws of mathematics or logic alone. They are normally singular or existential statements. They are observational in the sense that they can be compared directly with empirical evidence which is the immediate result of observation. They are relatively few in number or at any rate not infinite, for they correspond to actual observations which the scientist has made or will make.

3. *The scientist makes the indicated observations and compares their outcomes with the derived observational consequences*. The outcome is a decision for each prediction of whether it is consistent with or in conflict with observation.

4. *If all the derived observational consequences are confirmed by observation, the scientist regards the theory as more plausible than before. If some are in conflict with observation he rejects or modifies the theory*. In the latter case he knows the theory must be false by the Modus Tollens rule of logic. In the former case there is no parallel justification for thinking the theory is true, but it is said to be more fully 'confirmed' or 'corroborated' than before the observational evidence was examined.

According to Popper's account, science is a battlefield in which all theories are tentative and provisional conjectures in constant danger of being shot down by new observations. Theories which have stood up the longest under heaviest fire are to be preferred, but no theory is ever completely secure.

This account is oversimplified and has to be interpreted leniently. The scientist may not even formulate his theory until after he has made his observations. There may be many scientists involved, and the method may even be more of a social or historic process than an individual activity. The falsification of a single observational consequence often does not result in the rejection of a theory but instead in the reinterpretation of the observation;

or, the theory may be kept in the knowledge that it is false because all other known theories conflict with even more observations. But after all these qualifications are made and more, it can still be said that the hypothetico-deductive account has a germ of truth in it as a normative account in the context of justification.

These ideas apply directly to language descriptions: *A language description is a scientific theory about the language it purports to describe*. It is a statement or set of statements, preferably in a precise notation, about the subject matter of interest to the language scientist – the verbal behavior of the language users. To test the description the linguist must derive observational consequences from it and compare them against observed verbal behavior. If there is a fit, he grows more confident of the descriptive accuracy of his language description; if not he rejects or alters it. No finite amount of testing will ever make him absolutely certain it is correct, for there is always the possibility that some utterance form or speech situation not yet considered might not be accurately predicted by the description. But after corroboration by a sufficiently extensive body of observational or experimental data he may become reasonably confident of its approximate accuracy.

Some methodologists distinguish between a theory *per se* and its observational consequences, implying there may be more to a theory than the sum of its observational parts. The most striking difference between a theory and the sum of its observational consequences is, according to this view, that the theory statement may make use of *theoretical expressions* that don't denote anything observable, while the observational consequence statements are made up exclusively of expressions with direct empirical content. Thus a theory may talk about both 'observables' and 'non-observables' while the observational consequences are confined to 'observables'. It is not entirely clear how the theoretical expressions are to be distinguished from other expressions, and it is even less clear what the 'non-observables' are that the theoretical expressions denote, but there seems to be general agreement that some distinction along these lines is in order.

Structural language descriptions make heavy use of theoretical expressions. A structural language description specifies a particular language automaton; the structural features of the internal mechanisms of this automaton are the 'non-observables', its input–output actions the 'observables'. Everything in the language description which mentions the non-observables is a theoretical expression. Or to characterize them another way, all portions of the description which do not state behavioral properties involve theoretical expressions. Behavioral language descriptions, on the other hand, avoid the use of theoretical expressions entirely since by definition they state nothing but behavioral properties.

The final line of a structural language description (e.g. line (5) in Example 3.2, p. 48) has a less conventional status. It is best viewed, in the light of the hypothetico-deductive analysis of the scientific method, as an addendum to the theory indicating what is to be regarded as observable and what is not. According to traditional accounts of the hypothetico-deductive method, directions for identifying the theoretical expressions are not included in the formal statement of the theory itself. Matters of observability vs. non-observability are merely understood, or are at best included in informal auxiliary remarks on how the theory is supposed to be applied. But in the present development structural language descriptions go a step beyond this to formalize the observable/non-observable distinction as an integral part of the theory statement itself. The last line of the description tells all that the language scientist really needs to know about this distinction, namely that in investigating the object of ultimate interest he is supposed to abstract from all structural considerations and regard as empirically significant only those properties common to all members of the equivalence class, namely the behavioral properties.

As the hypothetico-deductive method is usually described, the observational consequences of a theory are supposed to be *directly* compared against what is actually observed, and there is supposed to be no ambiguity about whether the evidence is in accord with them or not. In practice, however, many scientific theories are simply not like that: the act of comparing the predictions of a theory against what is actually seen and heard is not a trivial or purely mechanical operation. This is especially true in the behavioral sciences. Some writers indicate this general problem area by speaking of 'indirect observables' in connection with entities that are not exactly imaginary, but which are nevertheless not immediately available as sense-data either. For example, Abraham Kaplan writes (1964, p. 55):

> *Indirect observables* are terms whose application calls for relatively more subtle, complex, or indirect observations, in which inferences play an acknowledged part. Such inferences concern presumed connections, usually causal, between what is directly observed and what the term signifies.... Such terms have also been called 'illata' and 'genotypes', and even, in some usages, 'hypothetical constructs'. Hypothetical they are, but there is nothing constructed about their referents. We infer their existence, but if the inference is justified, they are as much a part of the furniture of the world as are the trappings named by observable terms.

A language description, i.e. a delineation of the input–output behavior of a particular class of language automata, has to deal in entities of this troublesome type. The inputs and outputs of language automata involve sentences, and also such indications of belief values as the 'Yes' or 'No' output signals of the automata telling whether a test sentence is currently believed or not. An argument can be made that the sentences, even in their surface structure manifestation, are only indirectly observable, since a certain amount of inference may be needed even to compare a physical sentence-token against an abstract sentence-type. But the problem is more obvious in connection with belief values, for which there is no directly observable physical manifestation at all. Suppose for example that a language scientist 'observes' at a certain time that a certain language user believes a certain sentence. What does this mean? Actually, the scientist cannot 'observe' in any direct sense the belief value which the user attaches to the sentence. Rather, what he has noted is that the language user says he believes it, or acts as though he believes it, or is willing to lay bets on it, or in some other way gives strong evidence of believing it. On the strength of this evidence the scientist assumes that he believes it, and requires of his mathematical model, the language automaton, that it yield the 'Yes' response when the sentence is submitted to it as a test input in a comparable state. Thus the believed/nonbelieved output signals, and similarly the implicit 'accepted' status of each incoming informative sentence, would seem to be best treated as indirect observables.

The important point, though, is that within the present theory all input–output is treated as observable in *some* sense, whether it be directly or indirectly. This is in contrast to matters of internal organization within the language automaton, which are regarded in the theory as totally unobservable. The practical import of the distinction is that the theory commits the language scientist to the task of finding out, as part of the evidence-gathering process, what particular sentences and belief values have occurred in an observed speech situation, whereas it does not commit him ever to find out empirically anything about, say, the exact form or structure of the information states the speakers are in. All of this is in reasonable accord with common sense, for an intelligent language user is generally able to say something about whether he believes a given sentence or not, but the issue of how the totality of his information is 'structured' might not even be a meaningful question to him.

In summary, a language description can be regarded as a conventional scientific theory whose subject matter happens to be how some language is used to communicate. The only novel feature of a language description as a scientific theory is that if it is a structural description, an explicit statement indicating in effect which expressions are theoretical is included as a proper part of the theory statement. This is made possible by the formalization of the theory within automata theory, which as the theory of black-box analysis incorporates the distinction between observables and non-observables as an integral part of its formal apparatus. Aside from this departure, which we regard as an advantage, whatever can be said about scientific investigation in general applies to the formulation and testing of language descriptions in particular.

3.4 Basic evidence properties

The descriptive accuracy of a language description is tested by deriving observational consequences from it and comparing them against empirical data. What sort of observational consequences should be derived for this purpose? For a start, it is obvious that any observational consequence must be a behavioral property. In addition it seems natural to require that they be 'specific' properties – properties that assert something quite detailed about a narrow facet of the language. That they should not be too broad becomes apparent when one thinks of the limiting case of a property so broad as to characterize the entire language: such a property would be an entire language description rather than a consequence of it.

An important class of observational consequences which fill these requirements are what we shall call *basic evidence statements*. Roughly speaking, a basic evidence statement is an assertion of the existence or nonexistence of an information state with specified characteristics. Here is an example of a basic evidence statement:

(i) There exists an information state z such that *Jupiter is bigger than Mars* is believed in z.

In a more formal setting, this basic evidence statement would of course be recast as a mathematical statement affirming the existence in any language automaton for English of states such that the output in response to the test input *Jupiter is bigger than Mars* would be 'Yes'. If this property were derived from a formal description of English, one would be inclined to say what it claims is in accord with observed usage, for the possibility of possessing information that would cause one to believe the sentence in question is amply demonstrated by the existence of the many English speakers who actually do believe it. Thus the derivation of this basic evidence statement and its agreement with observation would tend to verify or corroborate to some extent the adequacy of the description of English from which it was derived.

Other basic evidence statements of the same form may be constructed by substituting other declarative English sentences in the same schema. For example,

(ii) There exists an information state z such that *Mars is bigger than Jupiter* would be believed in z

is another basic evidence statement that might also be derived from a description of English. It too would tend to corroborate the language description, for though we as language investigators may happen personally to disbelieve *Mars is bigger than Jupiter* there would be nothing to prevent a rational but misinformed English speaker from believing it.

A slightly more complex type of basic evidence statement is one involving a conjunction of belief assertions, as for example:

(iii) There exists an information state z such that *Jupiter is bigger than Mars* is believed in z and *Mars is bigger than the earth* is believed in z.

English would seem to possess this property too, for there is certainly nothing impossible about believing both the sentences it mentions.

Here, however, is a basic evidence statement of the same general form which does *not* accurately describe English usage:

(iv) *There exists an information state z such that *Jupiter is bigger than Mars* is believed in z and *Mars is bigger than Jupiter* is believed in z.

(The asterisk preceding the statement of the property indicates that the property is not an actual one, and is being mentioned only hypostatically. This notation is an extension to the pragmatic level of the standard linguistic device used to indicate the nongrammaticality of faulty sentence forms.) Statement (iv) is not in accord with English because no English speaker, whatever his factual information, could rationally believe both sentences simultaneously. If someone did claim to believe them both, one would have to challenge either his understanding of English, particularly his understanding of the full meaning of the comparative construction, or else his intellectual capacity for applying his linguistic knowledge accurately in this particular situation. A language description of which (iv) is a consequence is therefore not an accurate description of English. The opposite basic evidence statement, which is the same as (iv) except that it starts out "There does not exist a z . . . .", would however be in accord with the facts; if derived from a putative description of English it would tend to support the adequacy of the description.

Another kind of basic evidence statement is one involving the learning of one or more sentences, as well as an assertion of belief or nonbelief. It is true of English, for example, that

(v) There exists an information state z such that, upon learning *Jupiter is bigger than Mars* and then *Saturn is bigger than Mars*, z believes *Mars is bigger than the earth*.

Someone who already believed the third statement even before learning the first two would exemplify a language user in such a state. On the other hand, it is not true of English that

(vi) *There exists an information state z such that, upon learning *Jupiter is bigger than Mars*, z believes *Mars is bigger than Jupiter*.

so it must be marked with an asterisk.

The most general form of basic evidence statement is not only conjunctive like (iii) and (iv) but also involves learned sentences like (v) and (vi). An example would be:

(vii) There exists an information state z such that upon learning *Mars is bigger than Jupiter* z would believe *Mars is bigger than the earth*, whereas upon learning instead *Venus is bigger than Mars* z would believe *The earth is bigger than Mars*.

English has this basic evidence property since an English speaker knowing Jupiter to be bigger than the earth and Venus to be smaller, but who had no preconceptions about the size of Mars, would be in an information state satisfying it. More complicated basic evidence statements of the same general form can be constructed by introducing more sentences.

A *basic evidence statement* in a sentence set S is a metalanguage assertion of the general form

$$\text{There exists a (no) } z \in Z \text{ such that}$$

$$\bar{B}(z, \bar{s}_1) \text{ and } \ldots \text{ and } \bar{B}(z, \bar{s}_m)$$

$$\text{and not } \bar{B}(z, \bar{s}_{m+1}) \text{ and not } \ldots \text{ and not } \bar{B}(z, \bar{s}_n)$$

where $\bar{s}_1, \ldots \bar{s}_n$ are finite non-null sequences of members of S and $0 \leqslant m \leqslant n$ and $n \geqslant 1$. It is readily verified that every basic evidence statement expresses a behavioral property. (To accommodate incomplete languages each conjunct within a basic evidence statement may be expanded to read either (i) "$\bar{B}$ is defined for z and $\bar{s}_i$ and $\bar{B}(z, \bar{s}_i)$", or (ii) "$\bar{B}$ is defined for z and $\bar{s}_i$ and not $\bar{B}(z, \bar{s}_i)$", or (iii) $\bar{B}$ is not defined for z and $\bar{s}_i$.)

The basic evidence statements are so named because collectively they come as close as any class of evidence statements could come to providing a 'basis' for language description. We say a class of statements expressing behavioral properties constitutes a *basis* for language investigation if the properties of any given language could be determined by investigating empirically the truth of statements from this class alone. Then *each language is determined by the basis statements that hold true of it* (plus statements determining its sentence set). The members of a basis are like questions on a standard questionnaire which, when answered for a language, are sufficient to determine what the language is. It is convenient to know of a basis for language investigation, for it assures the investigator that all the evidence he needs about a language can be gathered by investigating properties in the basis. The basic evidence statements form a basis for the investigation of languages with finite structuralizations, and they 'almost' form a basis for all other languages, too, in a sense to be discussed.

Formally, a *basis* for the analysis of a family $\mathcal{F}$ of languages all with sentence set S is a set of behavioral properties (or metalanguage statements expressing behavioral properties) such that for any two distinct languages in $\mathcal{F}$ there exists in the set a property possessed by one but not the other. A language is said to be *finitely structuralizable* if and only if it has a structuralization $\langle S, Z, L, B\rangle$ in which S and Z are finite. In order to construct a basis for a family $\mathcal{F}$ of finitely structuralizable languages with given sentence set S, we consider basic evidence statements of the following sort. A basic evidence statement is *k-exhaustive* in a finite sentence set S if and only if it contains all possible expressions of form '$\bar{B}(z, \bar{s})$' where $\bar{s}$ is a non-null sequence of members of S of length no greater than k where $0 < k < \infty$. Here for example is a 2-exhaustive basic evidence statement in the sentence set $S = \{s_1, s_2\}$:

> There exists a $z \in Z$ such that $\bar{B}(z, \langle s_1\rangle)$ and not $\bar{B}(z, \langle s_2\rangle)$ and $\bar{B}(z, \langle s_1, s_1\rangle)$ and $\bar{B}(z, \langle s_2, s_1\rangle)$ and not $\bar{B}(z, \langle s_1, s_2\rangle)$ and $\bar{B}(z, \langle s_2, s_2\rangle)$.

Defining the *state-size* of a finitely structuralizable language to be the number of states in one of its reduced structuralizations, the result of interest may be expressed thus:

THEOREM 3.1. Let S be any finite sentence set and k any positive natural number. Then the (finite) set of all basic evidence statements that are k-exhaustive in S is a basis for the family $\mathfrak{F}$ of all languages whose sentence set is S and whose state-size is no greater than k.

Proof. The proof uses as a lemma a well-known theorem of automata theory due originally to Edward F. Moore (1956, p. 145). Translated into present terminology it states that in a (complete) finite language automaton with no more than k states, if two states are accurately 'described' by the same $(k-1)$-exhaustive basic evidence statements of existence form, then they are state-equivalent. (For simplicity, language automata that are not complete will be left out of consideration here.)

Let $\mathfrak{A}_1$ and $\mathfrak{A}_2$ be (complete) reduced language automata with sentence set S, each having at most k states and each satisfying the same k-exhaustive basic evidence statements in S. It will suffice to show that $\mathfrak{A}_1$ and $\mathfrak{A}_2$ are equivalent. This may be demonstrated by noting that both are isomorphic to a third language automaton $\mathfrak{A}_3$ constructed as follows. A k-exhaustive basic evidence statement will be said to be *in canonical form* if and only if it is an existence rather than a nonexistence statement and its conjuncts are ordered according to some predetermined rule, say by a lexicographic ordering of the sequences of sentences they involve (the latter requirement assures that no two distinct basic evidence statements in canonical form are ever equivalent). Let the state set Z_3 of $\mathfrak{A}_3$ be the set of all k-exhaustive basic evidence statements in S which are in canonical form and which are satisfied by $\mathfrak{A}_1$. Clearly each state in $\mathfrak{A}_1$ is 'described' by one and only one such metastatement. Moreover, by the lemma, no such metastatement 'describes' more than one state in $\mathfrak{A}_1$. Thus there is a one-to-one correspondence between Z_3 and Z_1. The belief relation B_3 of $\mathfrak{A}_3$ is constructed according to the requirement that $B_3(z, s)$ if and only if z contains as an unnegated conjunct the assertion that $\bar{B}(z, \bar{s}_i)$ where $\bar{s}_i = \langle s \rangle$. Clearly $B_3(z, s)$ if and only if $B_1(z^*, s)$, where z^* is the state in $\mathfrak{A}_1$ corresponding to z in $\mathfrak{A}_3$. The learning operation of $\mathfrak{A}_3$ is constructed by the rule that $L_3(z, s)$ is the metastatement z' in which the expressions $\bar{B}(z, \bar{s}_i')$ involving sequences $\bar{s}_i'$ of length $k-1$ or less occur negated or unnegated according as the corresponding expressions $\bar{B}(z, \bar{s}_i)$, where $\bar{s}_i$ is $\bar{s}_i$ preceded by s, occur negated or unnegated in z. The lemma assures that this rule specifies a unique z', and we have $(L_3(z, s))^* = L_1(z^*, s)$. Hence $\mathfrak{A}_3$ is isomorphic to $\mathfrak{A}_1$; and by the same reasoning $\mathfrak{A}_3$ is also isomorphic to $\mathfrak{A}_2$. It follows *a fortiori* that $\mathfrak{A}_1$ and $\mathfrak{A}_3$ are equivalent.

It is worth noting that this proof provides a method of constructing a definite structuralization for any finite language automaton already described behaviorally in terms of the k-exhaustive basic evidence statements it satisfies.

The theorem tells us that after a language investigator has determined the sentence set of a finitely structuralizable language he could in principle, provided he is able to set some upper bound on the language's state-size, write down a finite list of basic evidence statements, the empirical determination of whose truth or falsity would guarantee him sufficient data to formulate a fully detailed description of the language. In practice the inordinate length of this list and the statements on it would make such a procedure unworkable for languages of any substantial size, but the theorem nevertheless provides some assurance that no crucial features of a finitely structurablizable language will be systematically excluded if an empirical investigation of it is confined to testing out as many of its basic evidence properties as possible.

A closely related result which points to much the same moral is the following easy consequences of the foregoing theorem.

COROLLARY. If S is a finite sentence set, then the set of all basic evidence statements in S is a basis for the family of all finitely structuralizable languages with sentence set S.

Here there is no requirement that the investigator be able to set an upper bound on the state-size of the language he is analyzing. The price paid is that the basis, consisting of *all* basic evidence statements involving the sentences of the language under investigation, is not finite; hence there is no possibility even in principle of investigating separately every property in the basis. But the knowledge that the basic evidence statements form a basis is valuable nonetheless, for it gives assurance that by sampling properties taken exclusively from this basis one is not systematically excluding from the investigation some important general type of language property.

Few languages of practical interest are likely to be finitely structuralizable; certainly no natural languages are so. And when a language is not finitely structuralizable, no *finite* set of basic evidence statements is sufficient to determine it completely. Thus the basic evidence statements do not, strictly speaking, provide a complete basis for the analysis of the more interesting languages. Nevertheless, it seems reasonable to assume that they may at least approach that desideratum in the sense that their failure to form a basis is due only to 'problems of infinity', so to speak, rather than to any systematic exclusion of important varieties of observable linguistic data. If the language investigator were to test more and more basic evidence statements of greater and greater length, the presumption is that he would come closer and closer to a full determination of the language, approaching it 'asymptotically' as it were. This, coupled with the fact that it is reasonable to assume that there must be limits of some sort on the complexity of a useful language's rules, suggests the sufficiency of the basic evidence statements for most practical purposes, including evidence-gathering for languages that are not finitely structuralizable.

While this presumed 'asymptotic' sufficiency is not all that could be desired, no set of evidence statements could be expected to come much closer to providing a basis. The remaining problems stem from the quite well-established generalization that all language scientists are mortal, and so can verify observationally only a finite number of evidence statements all of which must be of finite length. Recall too that from the general perspective of the hypothetico-deductive method it is usually too much to hope that a theory will ever be proved with finality. The most that can be expected is that one's confidence in it will grow ever greater as more and more evidence is gathered, approaching a practical but never attaining an absolute certainty. This more modest hope the basic evidence statements seem quite adequate to fulfill.

3.5 THE EVIDENCE-GATHERING PROCESS

We have argued that the observational consequences derived from a language description to test its accuracy should, or may as well be, basic evidence statements. It remains to discuss the empirical techniques by which the truth or falsity of a basic evidence statement may be established. As in other sciences the means of empirical observation may be passive or active. If the approach is passive, the language scientist merely records unobtrusively the naturally-occurring usages germane to his theory. If it is active, he intervenes to bring about special conditions which would not otherwise have occurred. Active observation is another name for experimentation.

Linguistic experimentation includes such procedures as conversing with a native speaker of the language under investigation, having one native speaker make certain pre-arranged utterances in the presence of another native speaker, etc. One mode of experimentation that has proved especially fruitful in linguistic research is 'informant technique', which exploits the linguistic intuition of a cooperative native speaker, often the linguist himself.

It is to be expected that both passive observation and active experimentation, possibly including new kinds of informant technique, might be useful in gathering empirical data to test out basic evidence statements.

Of all the modes of data-gathering, passive observation is probably the least efficient; but it has its uses. The traditional 'corpus of utterances' upon which the description of a language's sentence set is sometimes based is in conventional linguistic analyses often obtained passively simply by recording normal speech as it naturally occurs, say, in casual conversations, or by examining pre-existing tests. The same kind of evidence might be useful in testing some of the simpler kinds of basic evidence properties. Suppose for example a native speaker of English were heard to utter the sentence *Jupiter is bigger than Mars*. If the speech situation is such that it is clear the speaker actually believes the sentence he has asserted (e.g. it is not a mere hypostatic usage) and the possibility of mental or verbal blunder seems remote, this observation could be taken as evidence that English possesses the basic evidence property that there exists an information state in which *Mars is bigger than Jupiter* is believed. More involved kinds of passive observation could lead to the confirmation or rejection of more complicated basic evidence properties.

Note however that if the investigator wishes to test a language for a basic evidence property involving pre-specified sentences, he is unlikely to come across any usages of them through passive observation. Then too, no amount of passive observation can ever either verify a basic evidence statement of the nonexistence type or falsify a basic evidence statement of the existence type. These drawbacks of passive observation in pragmatic analysis are parallel to its recognized drawbacks as a basis of syntactic analysis.

Hence active experimentation ordinarily offers the most hope. For instance, the linguist might simply ask a language user outright "Do you believe the sentence *Jupiter is bigger than Mars*?" If the subject says "Yes" the linguist would have evidence that there exists an information state in which that sentence is believed. This kind of experimentation allows basic evidence properties involving arbitrary sentences to be explored, but still does not overcome the problem of how to verify nonexistence statements or falsify existence statements. Failure to produce an information state in an experimental situation does not prove it could not exist.

On the syntactic level the analogous problem is met by employing an informant technique in which a native speaker of the language is asked something like "Can you think of any legitimate speech situations (excluding blunders and the like) in which _____ might be uttered?" If the informant says "No" the linguist has much stronger evidence against the grammaticality of the utterance than could be practically obtained in any other way. A similar informant technique would appear to solve the corresponding problem on the pragmatic level. Thus a native informant may be asked whether he can imagine circumstances, apart from blunders and the like, in which someone might be in an information state fulfilling the specifications of the basic evidence statement of interest. If the informant says "No" the investigator has evidence tending to verify the statement if it is of the nonexistence type and to falsify it if of the existence type. Consider for instance basic evidence statement (iv) on p. 57. The informant would be asked in essence whether it would be possible for a rational English speaker simultaneously to believe *Jupiter is bigger than Mars* and *Mars is bigger than Jupiter*. If he says "No", i.e. if he cannot conceive of this except through blunder, stupidity, non-standard use of English, etc., this would be direct evidence against (iv).

Informant technique of this sort is a form of gedanken experimentation. Physicists use the device of the thought experiment on occasions where everyone can agree what *would* happen if a certain experiment *were* to be performed. The technique of gedanken experimentation is ideally suited to linguistic data-gathering because in linguistics such occasions are the rule. Since every user of a language has an intimate knowledge of how the language is supposed to be used, he is able to predict how it would be used by any other native speaker adhering to its canons. To be sure, such informant technique has its risks, and its use on the pragmatic level may require a more thoughtful breed of informant than has been needed for syntactic investigation. But an intelligent informant has a remarkable capacity for thought experimentation in matters relating to his native tongue, and the prospect of testing at least some of the simpler basic evidence properties in this fashion seems far from hopeless.

Although in practice a linguist often acts as his own informant, it is important that the role of the informant be kept conceptually separate from that of the investigator. The informant should be called upon in his capacity as a native speaker who knows how his language is used in specific situations, and in no other capacity. In particular, any special preconceptions the informant may have about linguistic theory should not be allowed to contaminate the evidence he supplies; the theoretical precepts on which the language description is ultimately based should be the domain of the investigator, not the informant. It follows that *an informant should never be asked 'conceptual' questions about his language, nor should he be required to respond to questions that require 'conceptual' answers*. If for the sake of efficiency the investigator does use some abstract terminology in his questioning, the terminology should be carefully defined for the informant in observational terms, and pains should be taken to ensure that the informant is really interpreting the abstract terminology as directed. If this guide line is not adhered to, the resulting language description is in danger of describing not what the rules of the language actually are, but instead the informant's preconceptions and prejudices about what the rules ought to be. All past experience with informant technique indicates that the two can be very different.

The investigator cannot help following this rule if he thinks of his informant merely as a performer of gedanken experiments, and therein lies the value of the gedanken-experimenter conception of the informant's role. A directly observed experimental result cannot of itself contain preconceptions, and as long as the informant is just a laboratory assistant running experiments for the investigator there is little danger of results colored by the informant's theoretical preconceptions.

The What-Do-You-Know? game offers a vehicle for gedanken experimentation. To use it the language scientist teaches his informant the game; then, to test a basic evidence property, he describes to the informant the course of a hypothetical What-Do-You-Know? game with questions and answers selected to correspond with the property. The informant is then asked, "Could a native speaker of your language be in a belief state which would cause him to play the way the Answerer did in this game?" An affirmative answer means the information state described in the basic evidence statement exists, a negative answer that it doesn't, so far as the informant can judge by experimenting mentally with various possible funds of information.

EXAMPLE 3.3. *The What-Do-You-Know? Game for Basic Evidence Statement (vi).* To test basic evidence statement (vi) on p. 58 the investigator would allow his informant to study the following game protocol:

> Q: *Jupiter is bigger than Mars.*
> Q: DO YOU BELIEVE THAT *Mars is bigger than Jupiter?*
> A: Yes.

The informant would presumably say that the Answerer played incorrectly no matter what his original information, i.e., English does not have property (vi).

EXAMPLE 3.4. *The What-Do-You-Know? Games for Basic Evidence Statement (vii).* Statement (vii) on p. 58 requires that the informant be shown *two* game protocols:

GAME I

Q: *Mars is bigger than Jupiter.*
Q: DO YOU BELIEVE THAT *Mars is bigger than the earth?*
A: Yes. (End Game I.)

GAME II

Q: *Venus is bigger than Mars.*
Q: DO YOU BELIEVE THAT *the earth is bigger than Mars?*
A: Yes. (End Game II.)

After he has studied both games the informant is asked if an English Answerer could be so informed that if questioned as in Game I he would answer as in Game I but if questioned instead as in Game II he would answer as in Game II. If so, English has basic evidence property (vii).

Basic evidence statements which like (vii) call for the consideration of more than one What-Do-You-Know? game protocol at a time correspond to what in automata theory are called 'multiple experiments'. A *simple* experiment is performed on an automaton by feeding it a series of inputs and observing the resulting outputs. A *multiple* experiment is performed on an automaton by performing a simple experiment on it, then resetting it to whatever its initial state was beforehand and performing a different simple experiment, resetting it to that same state again and performing still another simple experiment, and so forth a finite number of time. Note that a multiple experiment is possible only if the automaton has, as it were, a 'reset button' on its side allowing it to be restored at any time to the state it was in before the experiment began. The distinction between simple and multiple experiments is known to be crucial for certain purposes, e.g. it is not always possible to deduce which initial state a reduced finite automaton of known design is in by performing a simple experiment on it, but it can always be deduced by means of an appropriate multiple experiment (Moore 1956).

Many basic evidence properties, e.g. (vii), clearly require the execution of what is essentially a multiple experiment for their empirical validation. Now a multiple experiment would be extremely awkward, if not impossible, to perform by any of the usual means of direct experimentation. The reason is that, inconsiderately enough, people do not come equipped with "reset buttons" by which they can be put back into a former belief state. Once the experimenter has caused a subject to learn a series of sentences, it would be very awkward for him to have to somehow make his subject 'unlearn' them, especially under the constraint that his methods must not destroy his credibility to the point where he cannot get the subject to learn a different series of sentences. With a gedanken technique, on the other hand, this problem does not arise. The informant is able simply to imagine himself or another speaker back into a former mental state. In the world of thought experiments, people *do* have reset buttons.

One has to conclude that there is a large class of basic evidence properties – namely, every basic evidence property to which there corresponds more than one What-Do-You-Know? game – which could hardly be investigated at all without the assistance of informant technique. Practically speaking the language scientist *must* rely on the imagination of an informant.

CHAPTER 4

LANGUAGE AND DEDUCTIVE LOGIC

We have seen how the idea of an information state leads to a way of analyzing languages on the pragmatic level. The role of logic in the analysis will be explored next. It will be suggested that a companion theory of logic does indeed exist provided suitable idealizations are presupposed.

4.1 IDEALIZATIONS

Whenever a mathematical theory is applied to observational phenomena, a host of assumptions is immediately invoked, usually unstated and sometimes subtle. Prominent among the assumptions are those about which aspects of the real world the abstract theory is supposed to account for. They are conceived sometimes as limits on the applicability of the theory, sometimes as idealizations introduced to keep the theory from getting out of hand. The idealizations can be more or less troublesome depending on the subject matter of the theory. Language science seems to be an area in which they are more troublesome.

Idealizations have of course a venerable place in the history of science. To use the well-worn examples, the frictionless pulleys and weightless ropes of classical mechanics are idealizations of what is to be found in the real world, as is the perfect vacuum and absence of force fields other than gravity assumed in the law of falling bodies. The use of such idealizations is not so much to assume away all annoying practical realities as it is to isolate them for separate scrutiny. In the case of pulley systems, for example, the idealization of frictionlessness allows the physicist to handle first the basic considerations of his problem that would be common to all similar pulley systems regardless of differences in coefficients of friction, and then to introduce the complicating factor of the friction into his considerations later if necessary as a refinement. In this way the idealization helps to separate underlying considerations of great generality out from the other complicating factors found in individual situations. But there is nothing mysterious about it; to idealize is simply to ignore certain side effects.

This commonplace view of the role of idealizations in science is certainly applicable to language science in particular, but in the case of language science there is more to it than just that. The ancients who first tried out the idealizations of classical mechanics had to make a strong working hypothesis, namely the assumption that the separation of influences applied by their idealizations was indeed a viable one. For all that was known beforehand, it might not have turned out that way. But in the case of language science it is clear from the very nature of communication that there has to exist something approaching a common core of agreed principles governing the usage of the code. Thus the very fact that successful language usage depends on shared conventions gives prior assurance that a language obeys laws that are 'ideal' in the sense that each language user must attempt to approximate them in his own use of the language.

It might be objected that the idealized rules to which each language user attempts to conform his own behavior vary somewhat from user to user because of differences of dialect and idiolect. The objection is valid but not fatal to the general conception. Even if each speaker is regarded as having his own internalized language rules that are a little different from those of every other speaker, it still seems clear that the rules must be fundamentally fairly similar for different speakers to be able to converse fluently with one another. It is as though each user attemps to conform his own usage to ideal rules which he hopes are close to the center of gravity of everyone else's ideal rules. This common desire to maintain conformity acts as a centripetal force holding the language together. If the centripetal force ever breaks down to the point where the different dialects are no longer mutually intelligible, there is by definition no longer just one language.

This fairly standard view of the situation has ramifications for language analysis at all levels. Consider first the syntactic level. To describe a language on the syntactic level one must specify rules delimiting the class of its well-formed sentences. Now, the rules of interest define what we will call the *ideal* sentence set of the language, not its *actual* sentence set. The ideal set conforms to the abstract norms of the language, the actual set contains all historically existing sentences and sentence attempts. Disclaimers appear repeatedly in the linguistic literature to the effect that it is not the actual sentence set that is typically of interest but the ideal. It is instructive to review some of the arguments establishing the difference between the two:

1. There exist members of the ideal sentence set that are not in the actual sentence set. This must be so because the ideal sentence set of a natural language is infinite but the actual set finite. That the ideal set is infinite may be demonstrated by exhibiting any indefinitely long series of sentences of increasing length; the successive stanzas of this folk song will do as an example:

She was looking back at me.
I was looking back to see if she was looking back at me.
She was looking back to see if I was looking back to see if she was looking back at me.

⋮

2. Among the infinity of sentences in the ideal set but not the actual, many could be uttered but did not and will not because of historical accident.

3. But even if the actual sentence set were extended to include all sentences that *could* physically be uttered, it would still not be as large as the ideal set. Many sentences in the ideal set are astronomically long, and others are so difficult (super tongue-twisters or mind-twisters as it were) that their utterance is as much a physical impossibility as a frictionless pulley.

4. To the objection that it is absurd to take astronomically long or impossibly difficult sentences seriously, the linguist has a counterargument based on the awkwardness of establishing a cutoff length. If lengths or difficulties are to be excluded due to physical unutterability, how is the maximum allowable length or difficulty to be decided? This question is indeed an awkward one, for no one knows what might be barely possible now or in the future.

5. If an objector persists in this objection, insisting that even though the line between the possible and impossible is hard to draw it must exist somewhere, the linguist still has as a final line of defense of his interest in the ideal set the following argument: Even if it could be established with finality that certain sentences are physically unutterable, it would remain true that if they *were* utterable they *would be* well-formed. This hypothetical knowledge is part of the native speaker's internalized model of the language, and its existence can be proved by producing native informants who will agree e.g. that the quadrillionth stanza of the above folk song is a good English sentence whether utterable or not.

6. As for the converse of arguments 1.–5., it is obvious that there exist members of the actual sentence set not in the ideal set. These consist of historic sentence attempts which for some reason were not or will not be well-formed. Examples include fragmentary utterances, false starts, speech blunders, and so on.

These standard arguments reveal much about what the ideal sentence set is commonly conceived to be.

The notion that a language is a set of shared conventions dates back at least to Aristotle and the Greek Conventionalists. The related distinction between ideal and actual utterances is usually credited to de Saussure (1916) who used the terms 'langue' and 'parole' in a comparable way. The notions of 'competence' and 'performance' introduced by Chomsky are also closely related, the passages in which they most resemble the present ideal-actual distinction being Chomsky's (1967, p. 398) and (1976, p. 9).

How does this syntax-oriented view of the ideal sentence set fit in on the pragmatic level? Rather well as far as it goes. The sentence set of a language automaton is just the idealized sentence set studied by linguists investigating syntax. But there is more to a language automaton than its sentence set, and the idealizations needed for information states are trickier because information states do not have a directly observable structure. We must distinguish between an *ideal* information state set and an *actual* one, the former being the set of all states an ideal language user is capable of being in and the latter containing all the historically existing states that the language users have or will be in. The two sets differ for reasons analogous to those cited for the syntactic level:

1′. There exist members of the ideal state set not in the actual state set. As before, the argument is that for natural languages the former is infinite and the latter finite. That the ideal state set for English must be infinite is seen by referring to the folk song once again: someone could believe the first stanza but not the second or following stanzas; or the second but not the third or following; and so on *ad infinitum*. (Caution: this argument does not prove that any one fund of information is infinite, any more than its counterpart on the syntactic level proved the existence of infinitely long sentences.)

2′. Among the infinity of information states in the ideal state set, many are never realized in a human mind because of the historical accident that no one ever has had or will have occasion to believe all and only the information in the fund which the state represents.

3′. But even if the actual information state set were extended to include all 'knowable' funds of information whether historically existing or not, it would still not be as large as the ideal state set. For there are states in the ideal set containing so much information that it is unthinkable that they could ever be realized in a human mind. There are others whose information is too difficult or complex to be grasped humanly. Thus the ideal set contains stores of information that it would be impossible for a language user to possess, just as classical mechanics talks about entities that are impossible.

4′. The proposal that the state set postulated for language analysis ought to contain only realistic states meets with the awkwardness of establishing an upper limit on the amount and complexity of the information that ought to be allowed in a realistic state. Any proposed limit would be highly arbitrary.

5′. And even if a reasonable limit could be established, the fact would still remain that funds exceeding it would still be of hypothetical interest. One can conceive of an ideal speaker being so informed as to believe the quadrillionth stanza of the folk song even if it is hopelessly beyond human intellectual capacity. The theorist should therefore be as willing to admit the existence of the state as the existence of the stanza.

6′. Conversely there are states in the actual state set that are not in the ideal state set. These are states which are malformed in some way, as for instance a state would be if it allowed the simultaneous belief of *Jupiter is bigger than Mars* and *Mars is bigger than Jupiter*. (One could of course blame the belief relation rather than the state for this anomaly; actually the ideal nature of the state set should, strictly speaking, be discussed only in conjunction with the belief relation and learning operation.)

From these arguments it would appear that the idealizations appropriate to pragmatic language analysis are a natural extension of those which it is standard practice to assume on the syntactic level.

For a broader perspective on the extended idealizations it is enlightening to consider the distinction between *information processing* and *information transfer*. Information processing is what goes on inside an information-handling entity and consists of such operations as storing, transforming, and calculating with information. Information transfer is what happens when information is transmitted from one such entity to another, as from speaker to hearer. When language science is construed as the study of how language is used to communicate, the focus is on information transfer. But even to study information transfer it is necessary at least to recognize the existence of information processing too. The recognition comes in the statement of the idealizations which say in effect that to study linguistic information transfer in an orderly way it is expedient to assume all information processing carried out in language users' minds to be ideal. The language users are to be described under the pretense that their information is always fully and correctly processed, the processing being instantaneous. This is no stranger than weightless ropes or rigid bodies. In order to concentrate on languages exclusively in their capacities as information transfer systems, it is natural and expedient to abstract from all considerations of information processing.

4.2 Logical Relationships

Some very interesting interrelationships among the sentences of a language can be specified in terms of its language automata. Some sentences are related to others by such circumstances as that if the former are learned the latter must be believed; or that the acceptance of some guarantees a lack of belief in others; or that not all can be believed simultaneously; etc. When such relationships are examined closely, many appear to have a logical character, especially when considered in the light of the idealizations just discussed.

Suppose as a first example that for two sentences s_1 and s_2 of a language it is determined that in any information state of one of the language's structuralizations, if the information of s_1 is known, the information of s_2 is known, too. (In other words, every state in which s_1 when submitted as a test sentence results in a 'Yes' response is also a state in which s_2 when submitted as a test sentence results in a 'Yes' response.) What is to be said about this relationship between s_1 and s_2? Insofar as it is impossible to believe the first sentence without also believing the second, it would appear to be some kind of implication relationship: s_1 implies s_2 in some way or in some sense. Since this, like conventional logical relationships, is a relationship between sentences (as opposed to something more esoteric like states), it seems appropriate to call it tentatively a relationship of 'logical implication'. In other words a kind of logical implication is defined by the condition that s_1 *logically implies* s_2 in a language if and only if for each of the states in a structuralization of the language, if s_1 is believed in the state so is s_2.

An objection frequently raised against attempts to define implication in terms having to do with human confidence in linguistic assertions is that human reasoning is frail, true logic immutable. It often happens that someone will have confidence in one sentence but not in another which it implies simply because the logical connection is not obvious to him. But this objection overlooks the idealizations that are in force here. The objection would be valid for a definition of implication couched in terms of actual beliefs and an actual state set, but the present definition is cast in terms of language automata describing the ideal rather than the actual. Hence the objection simply does not apply. Similar remarks apply to the converse objection that people often think a logical implication relationship exists where it doesn't.

The proposed criterion extends readily to sets of premises rather than single premises, and by taking the premise set to be empty to a definition of logical validity or tautology. Logical equivalence and independence are definable in terms of implication in standard ways. Logical incompatability is definable as lack of an information state in which the sentences in question would be simultaneously believed, and as a special case logical self-contradiction is definable as unbelievability.

DEFINITION 4.1. S_1 *logically implies* s in a language $\mathfrak{L}$ (in symbols, $S_1 \models_{\mathfrak{L}} s$) if and only if for some (equivalently, every) structuralization $\langle S, Z, L, B\rangle \in \mathfrak{L}$, $S_1 \subseteq S$ and $s \in S$ and for every $z \in Z$, if $B(z, s_1)$ for all $s_1 \in S_1$ then $B(z, s)$.

When no confusion can arise the subscript can be dropped from '$\models_{\mathfrak{L}}$'. If S_1 is empty $S_1 \models s$ is abbreviated $\models s$ and read "*s is logically valid*". (*In* case $\mathfrak{L}$ is not λ-complete the defining condition is interpreted: "... if for every $s_1 \in S_1$ B is defined for z and s_1 and $B(z, s_1)$ holds then B is defined for z and s and $B(z, s)$ holds.)

About the symbolism: Frege (1879) introduced the symbol '⊢' which he explained as a 'judgement stroke' followed by a 'content stroke'. Later, following the usage of Kleene and Rosser, his symbol came into more general use as an indication of the theoremhood of the sentence named after it. When also preceded by the name of a set of sentences the expression has come to mean that the sentence following is logically derivable from the set of sentences preceding. The symbol '⊨', used by Kleene in 1956, has come to replace '⊢' in contexts in which the definition of logical implication has a semantic rather than a syntactic character. Against this background it seems natural to add a third horizontal stroke and write '$\models$' for logical implication on the pragmatic level.

We define $\mathcal{C}$ to be the *logical closure operation* of a language $\mathfrak{L}$ with sentence set S if and only if $\mathcal{C}$ is the function from the set of subsets of S into the same set such that for any $S_1 \subseteq S$,

$$\mathcal{C}(S_1) = \{s \in S \mid S_1 \models_{\mathfrak{L}} s\}.$$

The set $\mathcal{C}(S_1)$ is the *logical closure* of S_1 and a set equal to its logical closure is said to be *logically closed* (also a 'closed system', 'deductive system'). Two sets of sentences S_1 and S_2 are said to be *logically equivalent* in $\mathfrak{L}$ if and only if $\mathcal{C}(S_1) = \mathcal{C}(S_2)$; in particular two sentences s_1 and s_2 are logically equivalent if and only if $\{s_1\}$ and $\{s_2\}$ are logically equivalent. A set of sentences is *logically independent* in $\mathfrak{L}$ if and only if it is not logically equivalent in $\mathfrak{L}$ to any of its proper subsets.

DEFINITION 4.2. A set S_1 is *logically inconsistent* (or 'logically incompatible') in a language $\mathfrak{L}$ (symbolically $\not\models S_1$) if and only if for some (every) structuralization $\langle S, Z, L, B\rangle \in \mathfrak{L}$, $S_1 \subseteq S$ and for every $z \in Z$ there exists $s_1 \in S_1$ such that not $B(z, s_1)$.

(In the incomplete case the "not $B(z, s_1)$" is read "if B is defined for z and s_1 then not $B(z, s_1)$".) A sentence s is *logically self-contradictory* if and only if $\{s\}$ is logically inconsistent. A set of sentences is *logically consistent* in a language if and only if it is not logically inconsistent in the language.

The symbol '$\Vdash$' was used by Frege to indicate that what followed was judged not to be the case, the short vertical stroke being a 'negation-stroke'. The extension of his notation to indicate inconsistency on the semantic ('$\Vdash$') and pragmatic ('$\models$') levels seems natural.

EXAMPLE 4.1. In the language automaton of Figure 2.2 (p. 23), and hence also in the Detectives' Language described in Examples 3.1 and 3.2, *Coe is in* and *Coe is out* are logically inconsistent, as are *Doe is in* and *Doe is out*. However, no sentence logically implies any other, there are no logically valid sentences, and there are no self-contradictory sentences.

An important methodological point here is that neither logical implication nor any of the other logical relationships are primitive concepts. This differentiates the theory from developments of logic in which at least one relationship, say implication or tautologousness, is primitive and the rest defined in terms of it. Here the logical properties and relations are all defined in terms of information states, and this development of the logical theory in terms of a linguistic primitive is what justifies calling the combined logical and linguistic system of ideas a 'unified' theory of language and logic.

4.3 Properties of the Logical Relationships

The pragmatic logical inconsistency and implication relationships defined in terms of information states make some intuitive sense, so the next step is to examine their mathematical characteristics to see if they are what is expected. The following characteristics follow immediately from Definition 4.1.

THEOREM 4.1. Let $\mathfrak{L}$ be a language with sentence set S. Then for any $s \in S$, $S_1 \subseteq S$ and $S_2 \subseteq S$,

(i) if $s \in S_1$ then $S_1 \models_{\mathfrak{L}} s$;

(ii) if $S_1 \models_{\mathcal{L}} s_2$ for every $s_2 \in S_2$ and $S_2 \models_{\mathcal{L}} s$ then $S_1 \models_{\mathcal{L}} s$:

(iii) if $S_1 \subseteq S_2$ and $S_1 \models_{\mathcal{L}} s$ then $S_2 \models_{\mathcal{L}} s$.

The logical closure operation allows this theorem to be stated in a form that is in some ways more transparent:

COROLLARY. Let $\mathcal{C}$ be the logical closure operation of a language $\mathcal{L}$ with sentence set S. Then for any $S_1, S_2 \subseteq S$

(i′) $S_1 \subseteq \mathcal{C}(S_1)$;

(ii′) $\mathcal{C}(\mathcal{C}(S_1)) = \mathcal{C}(S_1)$;

(iii′) if $S_1 \subseteq S_2$ then $\mathcal{C}(S_1) \subseteq \mathcal{C}(S_2)$.

Any set operation with properties (i′)–(iii′) is a *closure operation* (Birkhoff 1948). The corollary therefore states that our pragmatic logical closure is indeed a closure operation in at least one widely accepted sense. From this observation many others follow immediately by virtue of the well-known properties of closure operations: an intersection of closed sets is closed, a union of closures of sets is included in the closure of the union of the sets, the universal set S is closed, the family of all closed sets forms a lattice under the inclusion relation, and so on. Such properties are familiar characteristics of all conventional logical consequence operations.

This shows that $\models$ has some of the mathematical properties expected of an implication relation, but there remains the question of whether it has all of them. Some evidence that it does have at least the essential ones is to be found by comparing its characteristics with those that have historically been associated with implication in systematic formulations of the methodology of the deductive sciences. In one classic account (Tarski 1956, originally published in 1930), sentencehood and the logical closure operation are taken as the only primitives and the axioms laid down for the latter are essentially just (i′)–(iii′). (Tarski's postulates also include the compactness assumption that every set of sentences implying a sentence has a finite subset implying it, but this assumption is not included in all accounts of deductive methodology and would be an unnecessary restriction on the generality of the present investigation.) More recent works confirm that logical implication's closure properties are still regarded as its basic mathematical properties. Apparently $\models$ has the mathematical characteristics logical implication is supposed to have.

Some further immediate consequences of Definitions 4.1 and 4.2 follow.

THEOREM 4.2. Let $\mathfrak{L}$ be a language with sentence set S. Then for any $s \in S$ and $S_1, S_2 \subseteq S$

(i) if $S_1 \vDash_{\mathfrak{L}} s_2$ for every $s_2 \in S_2$ then $\not\models_{\mathfrak{L}} S_1 \cup S_2$ if and only if $\not\models_{\mathfrak{L}} S_1$;

(ii) if $\not\models_{\mathfrak{L}} S_1$ then $S_1 \vDash s$;

(iii) not $\not\models_{\mathfrak{L}} 0$.

COROLLARY. Let $\mathcal{C}$ be the logical closure operation for a language $\mathfrak{L}$ with sentence set S. Then for any $S_1 \subseteq S$

(i′) $\not\models_{\mathfrak{L}} S_1$ if and only if $\not\models_{\mathfrak{L}} \mathcal{C}(S_1)$;

(ii′) if $\not\models_{\mathfrak{L}} S_1$ then $\mathcal{C}(S_1) = S$.

For the most part the property $\not\models$ has the usual properties of logical inconsistency. However, there is one commonly assumed property it fails to have. According to the definition of logical inconsistency due to E. Post, a set of sentences is logically inconsistent if and only if its logical closure is the set of all sentences. (ii′) gives only the 'only if' part of Post's criterion. The reason is that the Post criterion is inappropriate for certain languages in which there happen to be no logically inconsistent sets of sentences. Codes of that kind exist and a general theory of language and logic ought to be able to deal with them as they stand. For all other languages the Post criterion is appropriate and follows easily from Theorem 4.2.

4.4 LOGICS

Since every language has a logical implication relation and inconsistency property, and since the other commonly encountered logical relationships are definable in terms of these, it is natural to say that they determine the 'deductive logic' of the language. Thus the deductive system of a particular language is representable abstractly by its system of implication and inconsistency and the sentence set within which they apply. A deductive logic considered in isolation from any particular language is then a complex of this sort which is the deductive logic of at least one language.

DEFINITION 4.3.

(i) $\mathfrak{D}$ is the *(deductive) logic* of a language $\mathfrak{L}$ if and only if $\mathfrak{D}$ is the 3-tuple $\langle S, \models, \not\models \rangle$ where S is the sentence set, $\models$ the logical implication relation, and $\not\models$ the logical inconsistency property of $\mathfrak{L}$.
(ii) $\mathfrak{D}$ is a *(deductive) logic* if and only if there exists a language $\mathfrak{L}$ such that $\mathfrak{D}$ is the logic of $\mathfrak{L}$.

This characterizes a logic indirectly *via* the concept of a language. Logics can also be characterized directly as follows.

THEOREM 4.3. $\mathfrak{D}$ is a logic if and only if $\mathfrak{D}$ is a 3-tuple $\langle S, \models, \not\models \rangle$ such that S is a non-empty set of finite non-null sequences of elements of a finite set V (the 'vocabulary' of $\mathfrak{D}$) and $\models$ is a relation between subsets of S and members of S and $\not\models$ is a property of subsets of S and the following conditions hold for all $S_1, S_2 \subseteq S$:

(i) $S_1 \subseteq \mathcal{C}(S_1)$;

(ii) $\mathcal{C}(\mathcal{C}(S_1)) = \mathcal{C}(S_1)$;

(iii) if $S_1 \subseteq S_2$ then $\mathcal{C}(S_1) \subseteq \mathcal{C}(S_2)$;

(iv) $\not\models S_1$ if and only if $\not\models \mathcal{C}(S_1)$;

(v) if $\not\models S_1$ then $\mathcal{C}(S_1) = S$;

(vi) not $\not\models 0$;

where '$\mathcal{C}(S_i)$' abbreviates '$\{s \in S \mid S_i \models s\}$'.

Proof. The 'only if' direction is given by Theorems 4.1 and 4.2 and their corollaries. For the 'if' direction let S, $\models$ and $\not\models$ satisfy (i)–(vi); a language $\mathfrak{L}$ must be shown to exist whose logic is $\langle S, \models, \not\models \rangle$. Let $\mathfrak{A}$ be a language automaton with sentence set S. Let the state set of $\mathfrak{A}$ be

$$Z = \{S' \subseteq S \mid \mathcal{C}(S') = S' \text{ and not } \not\models S'\}$$

(i.e. the states of $\mathfrak{A}$ are the logically closed consistent sets of sentences of $\mathfrak{A}$). By (vi) and (iv) Z is non-empty. Let the belief relation B of $\mathfrak{A}$ be such that $B(z, s)$ if and only if $s \in z$. Let the learning operation L of $\mathfrak{A}$ be arbitrary. Let $\mathfrak{L}$ be the equivalence class of $\mathfrak{A}$. It is to be shown that

$$\langle S, \models, \not\models \rangle = \langle S, \models_{\mathfrak{L}}, \not\models_{\mathfrak{L}} \rangle.$$

To see why $\models = \models_{\mathfrak{L}}$, suppose $S' \models s$. Then by the construction of Z, for every $z \in Z$ if $S' \subseteq z$ then $s \in z$. Hence by the properties of B, $S' \models_{\mathfrak{L}} s$. For the converse suppose $S' \models_{\mathfrak{L}} s$. By the properties of B, for every $z \in Z$ if $S' \subseteq z$ then $s \in z$. If not $\not\models S'$ then $S' \models s$ by the construction of Z; if $\not\models S'$ then $S' \models s$ by (v).

To see that $\not\models = \not\models_{\mathfrak{L}}$, suppose $\models S'$. Then from (iii), (iv), and (v) it may be deduced that any superset of S' must also be inconsistent. It follows by the construction of Z that there is no superset of S' in Z. Hence $\not\models_{\mathfrak{L}} S'$ by the requirements on B. Conversely suppose $\not\models_{\mathfrak{L}} S'$. By the construction of B there is no superset of S' in Z and in particular $\mathcal{C}(S') \notin Z$. By the construction of Z this implies $\not\models \mathcal{C}(S')$ whence, by (iv), $\not\models S'$.

Every language has one and only one logic, but some logics are the logics of more than one language. This prompts the thought that the theory of language includes the theory of logic, but the theory of logic does not include quite all of the theory of language on the present pragmatic level of analysis. Logicians and mathematicians often refer casually to logical calculi as 'languages'. This usage is certainly in the spirit of the present theory, but it goes against the letter in that strictly speaking a logic is in general something less fully specified than a language.

4.5 INFORMATIVE LANGUAGES HAVE INCOMPLETE LOGICS

The formal definition of an information automaton does not exclude the possibility that an information automaton may have only one state. However, such automata have strange properties. A one-state information automaton can never be informed of anything new because its stored information never changes. Nor can it ever provide an observer with any information with surprise value, since its output is totally predictable. It must be regarded as a degenerate sort of information automaton incapable of storing any real information.

Now if a language automaton has only one state, it is to be regarded as a structuralization of an *uninformative language*, since no one's mind could ever be changed by anything said in that language. The speaker of such a language would be stuck in one state his whole life, so far as information conveyable by the language is concerned. An *informative language*, on the other hand, can be provisionally characterized by the requirement that all its structuralizations have at least two states. Thus informative languages contain at least some sentences capable of conveying factual information that somebody might not already know. Obviously, all natural languages are informative languages. Many formal or artificial languages are informative too, since it would be possible in principle to use them for the communication of factual information.

The informative/uninformative dichotomy is a linguistic distinction having to do with whether or not languages can be used for genuine communication. Might a corresponding logical distinction exist? It does and is known to logicians as 'logical completeness' (not to be confused with automata-theoretic completeness or the logician's 'semantic completeness'). A deductive system is *logically complete* when in that system every sentence is either logically valid or logically self-contradictory. It turns out that the completeness/incompleteness distinction in logic is the exact counterpart of informativeness/uninformativeness: a language is informative if and only if its logic is incomplete.

Intuitively an uninformative language is one structuralizable with only one information state. The idea can also be formalized by requiring that in every structuralization of an uninformative language, each state is state-equivalent to every other state. This is satisfactory for languages which are complete in the automata-theoretic sense of completeness. (But it raises technical difficulties in the incomplete case. In order to capture the intuitive notion of informativeness more faithfully in incomplete languages one must use a slightly weaker concept of state-equivalence than that of Definition 2.4. Call states z_1 and z_2 of a language automaton $\langle S, Z, L, B\rangle$ *virtually state-equivalent* if and only if for every finite sequence $\bar{s}$ of sentence in S, if $\bar{L}$ is defined for z_1 and $\bar{s}$ and also for z_2 and $\bar{s}$, then for every $s \in S$ the following equivalence obtains: B is defined for $\bar{L}(z_1, \bar{s})$ and s and $B(\bar{L}(z_1, \bar{s}), s)$ holds if and only if B is defined for $\bar{L}(z_2, \bar{s})$ and s and $B(\bar{L}(z_2, \bar{s}), s)$ holds. Virtual state-equivalence is not an equivalence relation but is implied by state-equivalence and reduces to it in the complete case.)

DEFINITION 4.4. A language $\mathfrak{L}$ is *informative* if and only if for some (equivalently, every) structuralization $\langle S, Z, L, B\rangle$ of $\mathfrak{L}$ there exist $z_1, z_2 \in Z$ which are not (virtually) state-equivalent. $\mathfrak{L}$ is *uninformative* if and only if it is not informative.

DEFINITION 4.5. A logic $\langle S, \models, \not\models\rangle$ is *logically complete* if and only if for every $s \in S$, either $\models s$ or $\not\models s$. A logic is *logically incomplete* if and only if it is not logically complete.

THEOREM 4.4. For every language $\mathfrak{L}$, $\mathfrak{L}$ is informative if and only if the logic of $\mathfrak{L}$ is logically incomplete.

Proof. For the 'if' part suppose the logic of $\mathfrak{L}$ is incomplete. Then by Definitions 4.1, 4.2, 4.5 there must exist in some structuralization $\langle S, Z, L, B\rangle$ of $\mathfrak{L}$ a sentence $s \in S$ and states $z_1, z_2 \in Z$ such that B is defined for z_1 and s and $B(z_1, s)$ and it is not the case that both B is defined for z_2 and s and that $B(z_2, s)$. But clearly z_1 and z_2 are not virtually state-equivalent, so $\mathfrak{L}$ is informative. For the 'only if' direction suppose $\mathfrak{L}$ is informative so that for some structuralization $\langle S, Z, L, B\rangle$ of $\mathfrak{L}$ there exist members of Z which are not virtually state-equivalent. Then there exist $z_1, z_2 \in Z$ for which there exists a sequence $\bar{s}$ in S and $s \in S$ such that $\bar{L}$ is defined for z_1 and $\bar{s}$ and for z_2 and $\bar{s}$ and B is defined for $\bar{L}(z_1, \bar{s})$ and s and $B(\bar{L}(z_1, \bar{s}), s)$. Because $B(\bar{L}(z_1, \bar{s}), s)$, not $\models \neg s$. Because either B is undefined for $\bar{L}(z_2, \bar{s})$ and s or it is defined but not $B(\bar{L}(z_2, \bar{s}), s)$, not $\models s$. Hence the logic of $\mathfrak{L}$ is incomplete.

Many of the deductive systems encountered in mathematics are logically complete. For instance, the theory of the arithmetic of the natural numbers, taken to include all statements valid in the domain of the natural numbers, is complete as it is usually formalized (Tarski, Mostowski and Robinson 1953, pp. 30, 31). The language of this theory includes logical connectives, individual variables and quantifiers, operations such as + and $\cdot$, predicates such as < and =, and individual constants 0, 1, 2, etc. The logical completeness of the system simply reflects the fact that all sentences built up from these elements are either correct or incorrect statements about the natural numbers. Now, considered as languages, systems like this have a peculiar status, for their logical completeness shows they must be uninformative. What then are mathematicians really doing when they appear to be communicating with each other in one of these languages?

A traditional answer is that the mathematicians are not really conveying any information in the ordinary sense, but are merely exercising their own and each other's linguistic muscles. They may be conveying facts about the language itself, but they are not exchanging any extralinguistic information. If they were already ideal speaker-hearers, there would be no occasion for them to try to upgrade their linguistic abilities in this way and no sending and receiving of these sentences would be observed. But they are not ideal, and the apparent uses of the language arise from the discrepancy between theoretical competence and their present performance capability. It might be thought a strange doctrine that '7,945 + 8,777 = 16,722' conveys no information; but what ordinary kind of information could this be that the hearer could figure out for himself given enough time?

This traditional answer may not be entirely correct, but to the extent that it is the present theory may help illuminate certain formerly problematic aspects of it. Consider this aspect: For what future occasions could the mathematicians be exercising their linguistic muscles? Part of the answer may be this: a language which is not itself informative may nevertheless be embedded in a larger language that is, and practice in the rules of the sub-language may carry over into the larger language. Consider for example the above-mentioned formal language of arithmetic extended to include such new constants as *John's age, the number of sheep in Jones' field*, and so forth. These are constants of a radically different sort from 0, 1, 2, etc. They denote quantities which are in some information states not yet known, not even to ideal speaker-hearers. They allow the language to be used for the conveyance of factual information via such nontautologies as *John's age* = 25, or *the number of sheep in Jones' field* < *the number of sheep in Smith's field*. Facility in the original restricted formal language carries over into this extended language.

Pushing this line of thought further, recall that language automata do not exist *in vacuo* but are part of a larger behavioral picture. This is why mathematical knowledge can influence nonverbal behavior as well as verbal. Even if there were no larger informative language to which an uninformative mathematical language could be extended, the mathematical language might still be worth learning because it had extensions in the language of action (cf. Figure 2.5). This could explain the deep-seated feeling that mathematical prowess is not entirely linguistic in character.

4.6 QUASI-LOGICAL RELATIONSHIPS

Logical implication and inconsistency as defined in Section 4.2 are static relationships in the sense that their defining criteria mention believing but not learning. The static criteria are plausible, but there also exist other interesting criteria of a dynamic character. For example, there is the condition: "For every information state a rational language user could be in, on learning s_1 his resulting new state would cause him to believe s_2." With the appropriate idealizations in force this sounds like an operational criterion for an implication relation of some kind; let us call it 'dynamic implication'. Dynamic criteria for inconsistency, equivalence, etc. are also readily formulated.

One hesitates to call dynamic implication, inconsistency, etc., 'logical' relationships for two reasons. First, the dynamic criteria make sense only for finite sets of premises or inconsistent sentences whereas the custom in mathematical logic is to regard logical implication and inconsistency as applicable to finite and infinite sets alike. Second, the dynamic criteria require the specification of a temporal order in which the premises are learned. A concern with temporal order is foreign to deductive logic as commonly conceived. As a compromise one might call relationships which are logical in flavor but which introduce a dynamic element 'quasi-logical'.

DEFINITION 4.6. $\bar{s}$ *dynamically implies* s in $\mathfrak{L}$ (abbreviated $\bar{s} \vec{\models} s$) if and only if for some (every) structuralization $\langle S, Z, L, B\rangle \in \mathfrak{L}$, $\bar{s}$ is a finite sequence of members of S and $s \in S$ and for every $z \in Z$, $B(\bar{L}(z, \bar{s}), s)$.

(For the incomplete case the last part reads "for every $z \in Z$, if $\bar{L}$ is defined for z and $\bar{s}$ and B is defined for $\bar{L}(z, \bar{s})$ and s then not $B(\bar{L}(z, \bar{s}), s)$".)

DEFINITION 4.7. s is *dynamically inconsistent* with s in $\mathfrak{L}$ ($\bar{s} \vec{\not\models} s$) if and only if for some (every) structuralization $\langle S, Z, L, B\rangle \in \mathfrak{L}$, $\bar{s}$ is a finite sequence of members of S and $s \in S$ and for every $z \in Z$, not $B(\bar{L}(z, \bar{s}), s)$.

(In the incomplete case: "... for every $z \in Z$, if $\bar{L}$ is defined for z and $\bar{s}$ and B is defined for $\bar{L}(z, \bar{s})$ and s the not $B(\bar{L}(z, \bar{s}), s)$".) When $\bar{s}$ is the null sequence dynamic implication and inconsistency reduce to logical validity and self-contradiction. Dynamic equivalence and other dynamic relationships are definable in terms of dynamic implication and inconsistency.

There are an infinity of other quasi-logical relationships which mix static and dynamic elements in various ways. To give an impression of what they are like, here are two further quasi-logical criteria for the assertion that s_1 and s_2 imply s_3:

(i) for every $z \in Z$, if $B(z, s_1)$ then $B(\bar{L}(z, s_2), s_3)$;

(ii) for every $z \in Z$, if $B(L(z, s_1), s_2)$ then $B(L(z, s_1), s_3)$.

When more than two premises are involved the number of possible defining criteria grows rapidly.

4.7 QUASI-LOGICAL RELATIONSHIPS ARE OFTEN LOGICAL

Under what conditions do the dynamic relationships coincide with their static counterparts? To be able to discuss this question methodically it is necessary to introduce some special language properties called 'stability', 'receptiveness', and 'intransigence'. 'Stability' is associated with the idea that if a sentence is already believed, nothing novel will be learned when it is heard or read. In such circumstances the hearer's or reader's information state should not change, and the languages whose structuralizations are like that are the stable languages. Unstable languages are probably best regarded as pathological and of no genuine interest.

DEFINITION 4.8. A language $\mathfrak{L}$ is *stable* if and only if for some (every) structuralization $\langle S, Z, L, B\rangle \in \mathfrak{L}$, for every $z \in Z, s \in S$, if $B(z, s)$ then $L(z, s) \simeq z$.

(In the incomplete case the defining condition is understood to mean that if B is defined for z and s and $B(z, s)$ holds, then L is defined for z and s and $L(z, s) \simeq z$.)

The second property, 'receptiveness', describes languages in which what is learned is immediately known. If someone hears or reads a sentence understandingly and accepts what it asserts, he finds himself in a state in which he believes the sentence. He does not instantly forget what he just learned, but sticks by it until such time as a further informative input does something to change his mind. Like stability, receptiveness is probably absent only in hypothetical languages of a pathological character.

DEFINITION 4.9. A language $\mathfrak{L}$ is *receptive* if and only if for some (every) structuralization $\langle S, Z, L, B\rangle \in \mathfrak{L}$, for every $z \in Z, s \in S, B(L(z, s), s)$.

(In the incomplete case the defining clause says that if L is defined for z and s then B is defined for $L(z, s)$ and s and $B(L(z, s), s)$.)

In stable and receptive languages there is no extensional difference between the logical and the dynamic sentence relationships for one-premise implications or two-sentence inconsistencies. This is convenient, for it means that when no more than two sentences are being compared the language analyst need not try to keep track of the static–dynamic distinction. Also, as a philosophical ramification of this situation, anyone who finds one of the two approaches to defining implication and inconsistency convincing and finds the stability and receptiveness properties natural has to concede the other approach too, when only a couple of sentences are involved.

THEOREM 4.5. Let $\mathfrak{L}$ be a stable and receptive language with sentence set S and let $s_1, s_2 \in S$. Then

(i) $s_1 \overrightarrow{\models} s_2$ if and only if $s_1 \models_{\mathfrak{L}} s_2$:

(ii) $s_1 \overrightarrow{\not\models} s_2$ if and only if $\not\models_{\mathfrak{L}} \{s_1, s_2\}$.

Proof. (The theorem will follow as a special case of the result to be proved next.)

'Intransigence' is the stubborn clinging to past beliefs. In an intransigent language a statement once learned can never be unlearned, or at least not as a result of further linguistic input. In contrast with stability and receptiveness, it is not at all clear that intransigence is a natural property to posit. Indeed, it is considered a mark of rationality to change one's mind on occasions when assertions newly received on sufficiently high authority call for it. However that may be, it is of interest that intransigence is the additional property needed to assure that static and dynamic relationships come to the same thing no matter how many sentences are involved.

DEFINITION 4.10. A language $\mathfrak{L}$ is *intransigent* if and only if for some (every) structuralization $\langle S, Z, L, B\rangle \in \mathfrak{L}$, for every $s_1, s_2 \in S$ and $z \in Z$, if $B(z, s_1)$ then $B(L(z, s_2), s_1)$.

(In the incomplete case the interpretation is "... if B is defined for z and s_1 and $B(z, s_1)$ holds and L is defined for z and s_2 then B is defined for $L(z, s_2)$ and s_1 and $B(L(z, s_2), s_1)$".)

THEOREM 4.6. Let $\mathfrak{L}$ be a stable, receptive, and intransigent language with sentence set S, let $s \in S$, let $\bar{s}$ be any finite sequence of members of S, and let S' be the set of all members of S occurring in $\bar{s}$. Then

(i) $\bar{s} \overrightarrow{\models}_{\mathfrak{L}} s$ if and only if $S' \models_{\mathfrak{L}} s$;

(ii) $\bar{s} \overrightarrow{\not\models}_{\mathfrak{L}} s$ if and only if $\not\models_{\mathfrak{L}} S' \cup \{s\}$.

Proof. (The generalization to the incomplete case is straightforward and will be left to the reader.) For the 'only if' part of (i), let $\mathfrak{A} = \langle S, Z, L, B\rangle$ be a stable, receptive, intransigent (and complete) language automaton and suppose $\bar{s}$ dynamically implies s in $\mathfrak{A}$'s equivalence class, i.e. for every $z \in Z$, $B(\bar{L}(z, \bar{s}), s)$. By repeated application of the stability assumption and the transitivity of $\simeq$ one can show by induction on the length of $\bar{s}$ that if for every $s' \in S' B(z, s')$ then $L(z, \bar{s}) \simeq z$, where S' is the set of all sentences occurring at least once in $\bar{s}$. This allows the substitution of z for $\bar{L}(z, \bar{s})$ in the original premise, giving that if for every $s' \in S'$ $B(z, s')$ then $B(z, s)$. That is, S' logically implies s in $\mathfrak{A}$'s equivalence class.

For the 'if' direction of (i), suppose S' logically implies s in the equivalence class for $\mathfrak{A}$, i.e. for every $z \in Z$ if for all $s' \in S'$, $B(z, s')$ then $B(z, s)$. By an induction on the length of $\bar{s}$ that uses receptivity and intransigence, it can be shown that for every $s' \in S'$, $B(\bar{L}(z, \bar{s}), s')$. Instantiating $\bar{L}(z, \bar{s})$ for z gives $B(\bar{L}(z, \bar{s}), s)$, i.e. $\bar{s}$ dynamically implies s in the equivalence class of $\mathfrak{A}$. The proof of (ii) is similar.

EXAMPLE 4.2. *Logical and Dynamic Relationships in the Detectives' Language.* The Detectives' Language is stable, receptive, and intransigent; hence by the theorem its logical and dynamic relationships are indistinguishable.

4.8 Logic in the Evidence-Gathering Process

The reader may have noticed that the defining clauses of the logical and quasi-logical relationships are all basic evidence statements. To assert that a certain logical or quasi-logical relationship holds among certain sentences of a language is therefore to assert that the language has a certain basic evidence property. Conversely, most of the simpler basic evidence properties are logical or quasi-logical relationships. For instance, of the seven illustrative basic evidence statements of Section 3.4 the first six affirm either the presence or absence of logical or dynamic relations. This brings out a significant point about the practice of pragmatic analysis: *much or most of a typical empirical investigation of a language is likely to consist of tests for instances of logical or quasi-logical relationships*. Whether recognized as such or not, tests for implication, inconsistency, equivalence, and so forth are the chief grist for the post-syntactic mill.

It follows that post-syntactic informant technique is largely a matter of getting the informant to render judgements about the validity of logical or quasi-logical relationships in specific instances. The informant is shown the sentences of interest and asked in effect whether or not a specified (quasi-) logical relationship holds among them. This does not mean of course that he is asked outright whether, say, one sentence is 'logically implied' by the others, for the informant may not know exactly what the investigator's definition of 'logically implied' is. The sounder practice would be to formulate the question without mentioning any logical terminology, say by posing it as a question about What-Do-You-Know? games as explained earlier.

For notational convenience, the device by which logical arguments are indicated in traditional logic provides a handy way of recording logical basic evidence properties. In an introductory logic textbook one is apt to find examples of logical arguments in natural language displayed thus:

(i) *All men are mortal*
Socrates is a man
———————
Socrates is mortal

This, the textbook will explain, is a valid logical implication with the sentences above the line its premises and those below its conclusion, the line itself indicating the deductive act. In the context of a pragmatic analysis of English of the present sort, for which the term 'logico-linguistic analysis' now seems justified, the line notation may be used to abbreviate the meta-statement that the sentences in question stand in the pragmatically-defined logical implication relationship. Thus (i) signifies that empirical tests have shown belief in the first two sentences to be accompanied invariably by belief in the third among rational English speakers.

The analysis may also borrow the traditional linguistic use of the asterisk and, employing it on the pragmatic level, write

(ii) *Some men are mortal*
Socrates is a man
*———————
Socrates is mortal

to record that this is *not* a valid implication under the pragmatic definition. The line-and-asterisk notation can clearly be used with any finite number of premises including none. Thus

———————
(iii) *All men are men*

records the finding that *All men are men* is logically valid, while

*———————
(iv) *Socrates is a man*

records that that sentence is not.

Logical inconsistency, whose pragmatic definition is like that of a conclusionless implication relation, may be recorded in this fashion:

(v) *All men are mortal*
Socrates is a man
Socrates is immortal
————————————.

Use of the asterisk converts this notation into a record of a finding of logical consistency, as in

(vi) *Jupiter is bigger than Mars*
Jupiter receives less solar radiation than Mars
* ————————————————————.

When only one sentence is involved the notation records self-contradiction or the lack of it:

(vii) *Some men are not men*
————————————

(viii) *Some men are philosophers*
* ————————————

A quasi-logical relationship can be recorded in like manner provided an indication is added of the kind of quasi-logical criterion used. A dynamic implication or inconsistency relationship, for instance, can be indicated by adding a vertical arrow:

(ix) *All men are mortal*
Socrates is a man
———————————— ↓
Socrates is mortal

The arrow is a reminder that the order of the premises is significant (though in this example they could be reversed with the same effect) and that they are understood to be learned in a top-down order.

When a traditional notation is used in a new way one must guard against falling into traditional thought patterns while using it. When in a traditional logic class the professor writes syllogism (i) on the board, he is apt to have in mind and impart subtly to his class the metaphysical assumption that there exists a Truth of Logic of which the syllogism is merely the linguistic expression. In the conventional wisdom it is conceived as a Truth of Aristotelian syllogistic, of set theory, of some other modern formal system, or as a supra-linguistic abstraction (it is not always clear which), but rarely as a simple pragmatic fact about English usage. The students see the validity of (i) because they glimpse the rationality of the Principle which it expresses, and the English is a temporary bridge to help them glimpse it; or so the traditional view goes.

There may be something to this view, but we would prefer to explain what is going on in rather different terms. What is really happening may instead be that the students are thinking like informants. The force of the English example to them is, by this account, that any clear thinker who has accepted the premises must also believe the conclusion, which is of course just the pragmatic definition of implication. Fortunately, in the Socrates example it does not matter whether the students think of the logical or the dynamic criterion or whether they confuse the two, for the example is an instance of both. And since in a traditional logic course they are never shown any examples in which it makes a difference which criterion is applied, they soon learn to lump logical and dynamic implication together in their minds and forget any distinctions between them that might originally have occurred to them. (English examples in which it *does* make a difference will be encountered in Chapter 8.)

What the students realize in the course of their gedanken experimentation with the Socrates example is in this view a fact about English. And this seems plausible, at least in the case of those students who at that point in the course have never been exposed to any other language. After learning one of the elegant formal languages they may, like their professor, come to think of the syllogism as expressing a more abstract Truth of Logic, but all that has actually happened is that they have seen some basic evidence properties of different languages that are roughly similar to one another. Perhaps there are no language-independent Truths of Logic, unless by that phrase is meant roughly comparable basic evidence properties held in common by all the languages with which one happens to be personally familiar.

It should be remembered too that the professor and the students are speakers of English. When one is a speaker of a language, especially a native speaker, it is hard not to think of the more general features of that language as absolute Truths, as a fish would think of the properties of water as logically necessary; and this is an instinctive tendency to be guarded against. In order to gain a more objective view of a language one must instead try to think like a Martian who has no idea what any of the human languages are like. The Martian cannot build on intuition and so is forced to carry out a genuinely scientific enterprise assuming nothing but what can be learned from observation of verbal behavior; he has the basic evidence properties as his starting point and nothing else. It seems likely that it is only by imagining what this whole enterprise would be like that we can hope to free ourselves from our comfortable insiders' mind-sets as native speakers and start to see our own languages and logics in the universal perspective of what *could* exist.

CHAPTER 5

SEMANTICS, AXIOMATICS

With the skeleton of a unified theory of language and logic now apparent, it is time to examine some of the possibilities for describing particular classes of languages within this framework. Particular hypotheses about what might go on 'inside the black box' are needed. It is at this point that many classical ideas about language and logic start to enter the picture, especially when the languages to be described have familiar classical characteristics. Among the classical ideas that can be used to flesh out the pragmatic skeleton for languages which call for them are model-theoretic semantics and traditional axiomatics. There may be languages to which neither approach is well suited; semantics and axiomatics are discussed only as *examples* of how a logico-linguistic analysis might proceed.

5.1 SEMANTICALLY STRUCTURALIZABLE LANGUAGES

In a structural language description, the language automaton which is described need have validity only up to automata-theoretic equivalence, so the analyst formulating such a description has complete freedom to postulate any kind of interior mechanism which results in the required input–output patterns. There are therefore many possibilities for describing the internal structure. It might be treated formally as a neural net, as a relay circuit, or as any number of other organic or electro-mechanical complexes. It could also be given in the form of a list of instructions for a specified computer. The internal mechanism can be presented as an abstract mathematical mechanism too – one for which there may be no obvious mechanical counterpart – and this approach has the advantage of not restricting horizons immediately to physically realizable structures (recall that because the rules to be embodied are ideal and valid only up to equivalence anyway there is no *a priori* reason to suppose they are physically realizable).

The first and most obvious mathematical mechanism to come to mind is to let information states be represented by sets of (believed) sentences. Unfortunately this technique is adequate only for languages of a restricted sort (cf. discussion Example 2.1). A more powerful representational technique adequate to describe this restricted class of languages plus many more is the 'semantic' technique now to be described.

Though it has not traditionally been viewed in that light, the body of theory known to metamathematicians as 'model theory' and to philosophers as one kind of 'semantics' offers a powerful way of constructing information states. This system of semantics (not to be confused with the 'general semantics' of Korzybski and Hayakawa, nor with recent proposals for semantical rules put forward by descriptive linguists) is usually credited to Alfred Tarski (1956, first presented in 1931) though many philosophers are more familiar with the variation of it presented by Rudolf Carnap (1942). The theory has been elaborated by Richard Martin, John Kemeny, Saul Kripke, and many others; some of the literature on it has been surveyed by Rogers (1963).

One way of approaching this kind of semantics is *via* Leibniz' notion of a 'possible world'. Just what this phrase means is not wholly clear, but for a start those who use it have in mind the universe, not just planet Earth. Moreover, since it is not the universe just at the present moment but throughout all time that is germane, a possible world must be understood as a possible universe-history. Everything that ever happened or will happen in the possible universe is included. 'Possible' means logically rather than physically possible; thus there are possible worlds in which objects fall upward, though there are none in which all objects fall upward and some do not.

Now consider the set of all possible worlds. For any fund of information someone might possess, some members of the set will be possibilities the information rules out while others will not; for instance the information presently in my possession rules out all possible universe-histories in which Gerald Ford won the American presidential election of 1976, but not all those in which he did not. This provides a conceptual basis for representing information: one tags each possible world with an indication of whether the given fund of information rules it out or not and the set of all possible worlds so tagged is a representation of the information in the fund. Another mode of representation which amounts to the same thing would be to take the set of all possible worlds which the information does not rule out as the information's representation.

The underlying assumption here is brought out by the following hypothetical procedure. Suppose a possible world could somehow be shown or described to someone in complete detail, after which he is asked whether for all he knows things really could be just that way. The procedure is repeated for another possible world and eventually (or so we try to imagine) such a judgement is elicited from him for every possible world. What is assumed in this way of representing information states is that his answers would characterize all of his current knowledge completely enough for purposes of language analysis. It is not assumed that the representation is adequate for analyzing all languages, but it is assumed adequate for some and for the present we restrict attention to these. The restriction to binary ruled-out/not-ruled-out decisions is arbitrary and will be given a probabilistic generalization later.

Conceived in terms of eliminated and uneliminated possible worlds, funds of information can be visualized as Venn diagrams. Figure 5.1(a) shows an information state represented in this way; the points in the shaded area are possible worlds that have been ruled out, the remainder are possible worlds not yet ruled out by anything currently known.

Let us now restrict attention further to languages in which, for any sentence and any possible world, one and only one of the following situations obtains: (1) the sentence would be true if the possible world were the actual world; (2) the sentence would be false if the possible world were the actual world. In case (1) we say the sentence *holds* in (or 'holds true of') the possible world; in case (2) we say it fails to hold in it. A sentence which holds in a possible world is also said to be *satisfied* by it; thus *Gerald Ford won the 1976 presidential election* is satisfied by all possible universe-histories in which he did indeed win that election. Philosophers sometimes speak of the satisfaction relation as embodying the 'truth conditions' of a sentence, meaning the rules determining whether the sentence would be regarded as true or false under specified circumstances. The restriction of attention is to languages for which it can be assumed that such rules exist for every sentence. Languages in which the notions of truth and falsehood are simply inapplicable for some sentences in some circumstances are to be left out of consideration at present.

Sentences can also be presented pictorially in Venn diagrams. In Figure 5.1(b) the sentence *s* is understood to hold in all the possible worlds within the circle and none outside; there is no shading. Sentences and information states can of course be represented in the same diagram as in Figure 5.1(c).

Consider now the inclusion relationship between state and sentence shown in Figure 5.2; what could be said about a language user in possession of such a fund of information? With a little thought it becomes clear he could be said to know or believe what *s* says, for the possibilities ruled out by *s* are among those already ruled out in his mind.

Next consider what would happen if a language user were to hear and accept a statement whose information content he had not known beforehand. If his prior knowledge was as shown in Figure 5.3(a), after accepting *s* his new state would be as shown in 5.3(b). In general, learning a sentence has the effect of ruling out all possibilities the sentence would rule out that are not already ruled out.

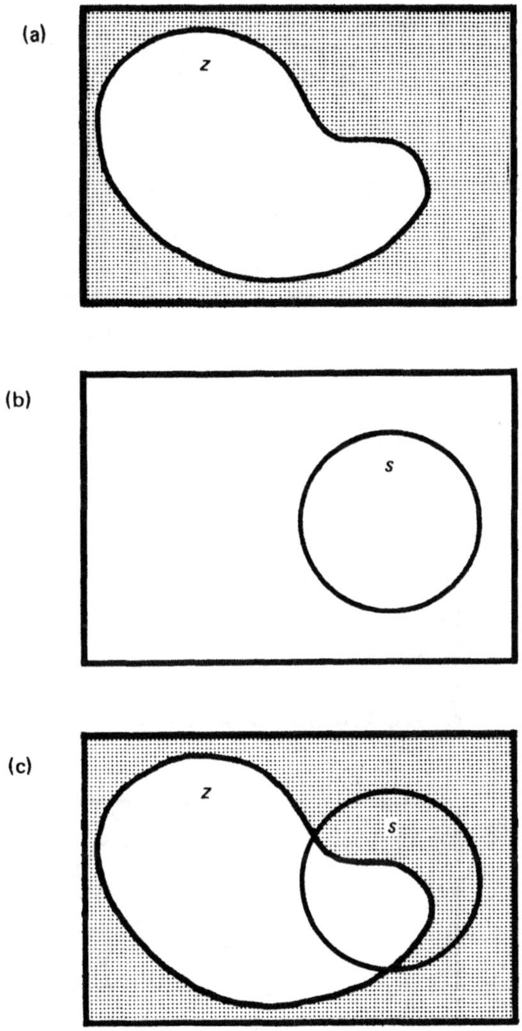

Figure 5.1. (a) A fund of information *z* represented diagrammatically by indicating which possibilities are 'ruled out'; (b) A sentence *s* represented as a set of possibilities of which the sentence would hold true; (c) A combined diagram showing a possible relationship of *s* to *z*.

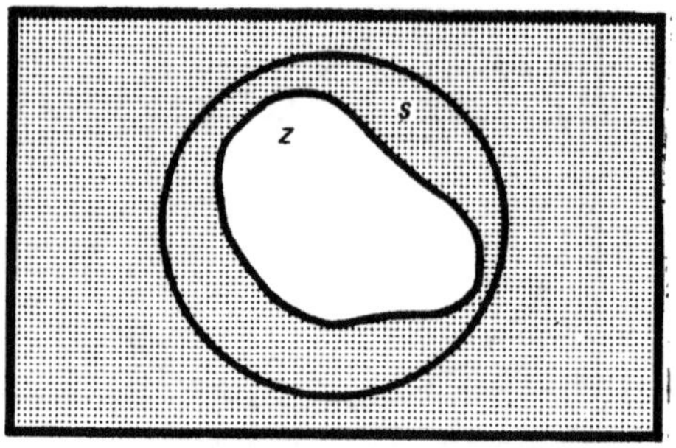

Figure 5.2. An information fund *z* and a sentence *s* such that a possessor of *z* would believe *s*.

(a)

(b)

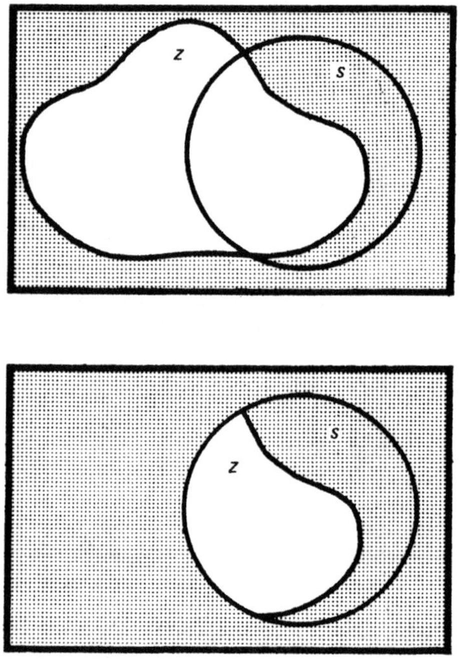

Figure 5.3. A fund of information *z* shown (a) before, and (b) after, a sentence *s* is heard and accepted.

We are now in a position to construct in imagination (though not yet in any precise mathematical sense) a conceptual device which is a sort of 'Leibnizian language automaton' for a language. Its sentence set is the ideal sentence set of the language. Each state in its state set is a set of possible worlds standing for uneliminated possibilities – that is, the state set is the set of all non-empty subsets of the set of all possible worlds. (Under the idealizations in force the empty set does not represent a fund of information a rational language user could have; it would not be self-consistent to think that *all* logical possibilities had been eliminated!) Its belief relation is the rule that a sentence is believed if and only if the set of possible worlds representing the current state is included in the set of possible worlds satisfying the sentence. Its learning operation is defined by the rule that the new state entered upon learning a sentence is just the set of possible worlds common to both the old state and the set of all possible worlds satisfying the sentence. An exception is the case in which this intersection is empty; in that case the learning operation is undefined.

Leibnizian language automata would surely be of importance if only it were possible to find viable mathematical representations for possible worlds. Unfortunately there is no hope whatever of finding such representations, because to be unique representations they would themselves have to be as complex as possible worlds. Such devices must therefore be consigned to the realm of conceptual motivation. However, it is sufficient for language-analytic purposes to construct a compromise structure which *behaves* just like a Leibnizian language automaton. Let us call two possible worlds *linguistically indistinguishable* in a language if and only if there exists no sentence in the language that holds true of one but not the other. In a Venn diagram indicating the borderlines of the areas corresponding to each of the sentences of the language, two points represent linguistically indistinguishable worlds if and only if no lines separate them.

Now, lumping linguistically indistinguishable worlds together could not affect the behavior of a Leibnizian automaton; to distinguish between linguistically indistinguishable possibilities is to draw finer distinctions than necessary. For language analysis it is therefore sufficient to deal with *indistinguishability classes* of possible worlds; one can use either the basic partitioning of the set of all possible worlds into indistinguishability classes or any convenient subpartitioning thereof. Indistinguishability classes may correspond to what some philosophers have called possible 'states-of-affairs'; they represent a possibility for what the world could be like but do not specify the possibility in full detail. The compromise Leibnizian language automaton has for its state set the collection of all non-empty sets of indistinguishability classes, and the satisfaction relation is specified with respect to the indistinguishability classes. It is worth noting that the retreat from possible worlds to indistinguishability classes of them allows some philosophically vexing questions to be bypassed for purposes of language analysis: these include all questions about what higher order of infinity of possible worlds there are and what degree of infinitesimal detail each includes.

The compromise automaton is more amenable to mathematical representation. To obtain a mathematical construct corresponding to the conceptual structure, any definite mathematical objects may be used to represent the indistinguishability classes so long as each class is assigned a distinct representation and the representations are such that the satisfaction relation can be specified conveniently in terms of them. For the familiar formal languages, ways are already known of representing the indistinguishability classes mathematically. They make use of structures involving expressions from the language under analysis plus standard set-theoretic entities. The results are called *modelling structures* (or 'models', 'interpretations').

Once the modelling structures and their accompanying satisfaction relation have been specified rigorously it becomes a straightforward matter to specify the rest of the desired language automaton in precise mathematical fashion. A language automaton constructed in this fashion will be said to be *semantically structured* and a language having at least one semantically structured structuralization will be called *semantically structuralizable*.

Let M be the set of all modelling structures and $H \subseteq S \times M$ the satisfaction relation. '$H(s, m)$' is read 's holds in m' or 'm satisfies s'. '$H[s]$' denotes the set $\{m \in M | H(s, m)\}$. Taking states to be sets of modelling structures interpretable as not-ruled-out indistinguishability classes of possible worlds, semantic structure is definable as follows.

DEFINITION 5.1.
(i) A language automaton $\mathfrak{A} = \langle S, Z, L, B \rangle$ is *semantically structured* if and only if there exists a non-empty set M (the 'model set' of $\mathfrak{A}$) and a relation $H \subseteq S \times M$ (the 'satisfaction relation' of $\mathfrak{A}$) such that

a) $Z = \{z | z \subseteq M \text{ and } z \neq 0\}$;

b) for every $z \in Z, s \in S$,

$$L(z, s) = \begin{cases} z \cap H[s] & \text{if } z \cap H[s] \neq 0, \\ \text{undefined} & \text{if } z \cap H[s] = 0; \end{cases}$$

c) for every $z \in Z, s \in S$,

$B(z, s)$ if and only if $z \subseteq H[s]$.

(ii) A language $\mathfrak{L}$ is ***semantically structuralizable*** if and only if some structuralization of $\mathfrak{L}$ is semantically structured.

From the definition it is seen that after its sentence set has been specified, the only hard part of constructing a semantically structured language automaton is specifying its M and H, for the rest of the automaton is immediately determined by these. To construct M and H it has been found convenient (for classical languages and a few others at any rate) to use as the main elements of the modelling structures mappings from the 'denoting' expressions of the language into appropriate set-theoretic entities. Modelling structures of this kind have been called 'relational structures', 'valuation functions', 'denotational interpretations', etc. A non-empty set is taken as the 'domain of discourse' of the structure, and the mappings in it associate elements of this domain with the proper names of the language, sets in the domain with predicates, sets of n-tuples with n-ary relation symbols, and so on. When there are whole sentences constructed as indecomposable symbols they are mapped into truth values, usually represented mathematically as 0 and 1. It has been found that modelling structures of this sort lead to simple statements of the semantical rules H in many classical languages.

Whether such modelling structures are adequate for the analysis of natural languages is problematic. The domain of discourse of a standard modelling structure is simply a set, representative presumably of some set of discrete real-world objects such as men or rocks. But the natural languages talk about many other kinds of things such as water, wealth, and democracy. Are the standard modelling techniques appropriate for such strange 'objects'? Or will it be necessary to elaborate them along radical lines, introducing, say, segments of the real line into the domain of discourse to represent the denotata of mass nouns? There is no way of knowing until some actual empirical analysis is undertaken, guided by rules of evidence clear enough to make it plain when one kind of structure succeeds better than another at predicting verbal behavior.

Tarskian/Carnapian semantics has been criticized on the philosophical grounds that the ontological status of the modelling structures, and especially of the elements in their domains of discourse, is unclear. One philosopher writes of the motivating idea of a set of all possible worlds that he is simply not prepared "to swallow this enormous metaphysical pill" (Ellis 1973,

p. 151). However, such objections lose much of their force when the system of semantics is viewed as part of a larger theory of pragmatics. In the present theory the possible world notion is not a metaphysical presupposition at all, but rather an idea suggestive of a black-box mechanism defining one possible class of languages. If such mechanisms fail to produce the behavior characteristic of the languages one is trying to describe, one rejects them as scientific hypotheses that do not fit the evidence, not as metaphysical postulates. The present use of the possible-world metaphor is separated from being an ontological assumption about reality by at least four removes: (i) it is not the proposed black-box mechanism itself, which is mathematical, but only a preliminary conceptual motivation for it; (ii) even the mathematical mechanism is not supposed to mirror reality except under strong idealizations; (iii) moreover it is supposed to have validity only up to automata-theoretic equivalence; and (iv) it is not claimed to be appropriate for all languages but only one logically possible kind of language whose resemblance to the natural languages remains to be investigated. The philosophical commitments involved in trying out a model-theoretic apparatus as part of a scientific black-box hypothesis are minimal indeed.

5.2 EXAMPLES OF ARTIFICIAL SEMANTICALLY STRUCTURALIZABLE LANGUAGES

Semantical rules are known for all the more familiar formal languages including propositional, first-order, and higher order systems of classical logic; intuitionistic logic; several systems of modal logic; and combinatory logic. When the semantical rules are known it is a simple matter to formulate a structural language description exploiting those rules within the pragmatic framework of the semantically structuralizable language family. Of course, the better-known formal languages were all invented for the purpose of formalizing mathematical proofs and for other metamathematical purposes, not for factual communication. It might seem odd for that reason to analyze them within a theory which seeks to explain language as a set of conventions for effecting information transfer. However, even if the familiar formal languages are not often used for the purpose of information transfer, nevertheless many *could* be so used given an appropriate interpretation. One could think of factual communication as the underlying capability of these languages which, even if unused, gives rise to the derivative logical properties which have been of interest to logicians and metamathematicians. Such languages offer the language analyst a chance to try a dry run of his descriptive apparatus without the need to gather evidence.

EXAMPLE 5.1. *A Simple Code*. As a first trivial example suppose the language to be described is a simple code, and that in this code there are only three possible messages 'P', 'Q', and 'R' of length one symbol each. When one user of the code wants to communicate with another he sends one of these three characters to him, printed on a card, say. It is assumed that each of the three symbols has some fixed interpretation known to all users of the code. For example, if the user population were a secret society of astronomers dedicated to discovering extra-terrestrial life before the rest of the world finds out about it, the symbols of the code might mean "There is life on Mars", "There is life on Venus", and "There is life on Jupiter" respectively.

It is assumed that the messages are logically independent of one another, e.g. that on learning P a user of the code would not be able to deduce Q on logical grounds alone.

The code can be analyzed as a semantically structuralizable language $\mathfrak{L}_1$. A structural description of $\mathfrak{L}_1$ proceeds by specifying the following semantically structural language automaton $\mathfrak{A} = \langle S, Z, L, B\rangle$. S contains just 'P', 'Q., and 'R'. The model set M of $\mathfrak{A}$ is the eight-member set of truth-value assignments in S, i.e. each modelling structure assigns either 'Truth' or 'Falsity' to each sentence in S. Z is the set of all non-empty subsets of M, and H is the rule that a sentence s holds in a modelling structure m if and only if s is assigned 'Truth' by m. L and B are determined from this in accordance with Definition 5.1 (ib, ic).

The formal language description looks like this (in lines (2a) and (3a) 0 and 1 represent falsity and truth):

(1) $S = \{P, Q, R\}$

(2)(a) $M = \{m | m: S \to \{0, 1\}\}$

(b) $Z = \{z | z \subseteq M \text{ and } z \neq 0\}$

(3)(a) For every $s \in S, m \in M$,

$H(s, m)$ if and only if $m(s) = 1$

(b) For every $z \in Z, s \in S$,

$$L(z, s) = \begin{cases} z \cap H[s] & \text{if this set is non-empty} \\ \text{undefined} & \text{otherwise} \end{cases}$$

(4) For every $z \in Z, s \in S$,

$B(z, s)$ if and only if $z \subseteq H[s]$

(5) $\mathfrak{L}_1 = [\langle S, Z, L, B\rangle]_{\cong}$

The format of the foregoing description could if desired be adopted as the standard format for all structural descriptions of semantically structuralizable languages. Note that lines (2b), (3b), (4), and (5) would remain the same in all such descriptions and so could be omitted. This would leave the specification of S, M, and H as the only essential parts of the description. In fact, so far as semantically structured language automata are concerned, we could if we wished replace the abbreviated 4-tuple representation of a language automaton by the still further abbreviated representation in the form of the 3-tuple $\langle S, M, H\rangle$.

EXAMPLE 5.2. *Contradictory Messages*. Now imagine another code $\mathfrak{L}_2$ whose sentences 'P', 'Q', and 'R' mean something like "There is life on Mars", "There is no life on Mars", and "There is life on Venus". The model set for $\mathfrak{L}_2$ must conform to the Venn diagram of Figure 5.4; that is, in any model in which P is assigned Truth Q must be assigned Falsity and *vice versa*. The formal description of $\mathfrak{L}_2$ is exactly the same as for $\mathfrak{L}_1$ except (2a) becomes

$$M = \{m | m: S \rightarrow \{0, 1\} \text{ and } m(P) \neq m(Q)\}.$$

Clearly, P and Q are logically and dynamically inconsistent in this language.

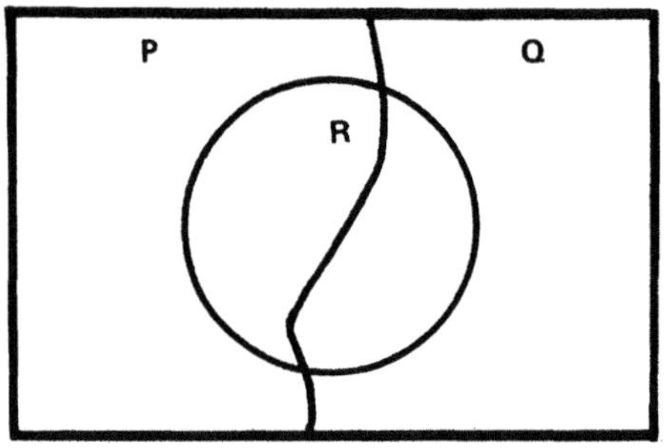

Figure 5.4. Venn diagram of model set of a semantically structured structuralization of $\mathfrak{L}_2$.

EXAMPLE 5.3. *A Different Language with the Same Logic*. Now imagine a language $\mathfrak{L}_3$ in which P, Q, and R mean "Venus is bigger than the earth", "Venus is smaller than the earth", and "Venus is either bigger and hotter or smaller and cooler than the earth". Because it is logically possible that Venus and earth are the same size, P and Q could be simultaneously false and the Venn diagram is as shown in Figure 5.5.

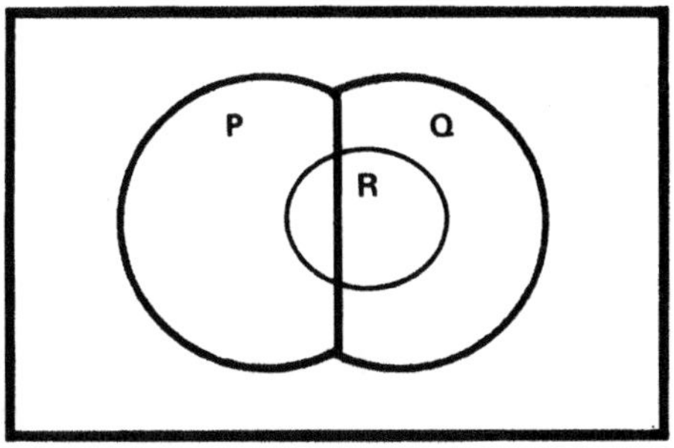

Figure 5.5. Venn diagram of model set for $\mathfrak{L}_3$.

It is noteworthy that the logic of $\mathfrak{L}_3$ is that of $\mathfrak{L}_2$; P and Q are inconsistent in both languages, etc. Yet $\mathfrak{L}_3$ is certainly a different language for it has a state (diagrammatically the area outside P and Q) in which the learning operation is not defined for either P or Q, and there is no such state in $\mathfrak{L}_2$. Evidently having the same logic does not guarantee being the same language even when both languages in question are semantically structuralizable.

EXAMPLE 5.4. *A Subject-Predicate Language with Classical Connectives.* Carnap and Bar-Hillel have described (Bar-Hillel 1964, pp. 221ff.) a little language adequate for expressing certain census information about a small community of only three inhabitants named 'a', 'b', and 'c'. The characteristics of interest in the census are maleness, femaleness, youngness, and oldness denoted by 'M', 'F', 'Y' and 'O' respectively. Any property symbol followed by an individual name (e.g. 'Ma') is a well-formed atomic sentence. In addition, if s_1 and s_2 are well-formed sentences so are '$\neg s_1$', '$[s_1 \,\&\, s_2]$', '$[s_1 \vee s_2]$', '$[s_1 \supset s_2]$' and '$[s_1 \equiv s_2]$'. These connectives are classical negation, conjunction, disjunction, (material) implication, and (material) equivalence respectively. With the help of the parentheses the connectives can be used to build up sentences of arbitrary length in the usual way (e.g. '$[Ma \,\&\, ((Mb \vee Yb) \supset Oa)]$' is a well-formed sentence).

The Carnap–Bar-Hillel language can be described logico-linguistically as a semantically structuralizable language $\mathfrak{L}_4$. The essential lines of the formal description follow.

(1) $I = \{a, b, c\}$

$P = \{M, F, Y, O\}$

$C_1 = \{\neg\}$

$$C_2 = \{\&, \vee, \supset, \equiv\}$$

$$S_0 = P + I$$

$$S_{i+1} = (C_1 + S_i) \cup (\{[\} + S_i + C_2 + S_i + \{]\}) \quad \text{for all } i > 0$$

$$S = \bigcup_{i=0}^{\infty} S_i$$

(2)(a) $M = \{\langle A, D\rangle | A \neq 0$ and Domain$(D) = I \cup P$ and for every $i \in I$, $D(i) \in A$ and for every $p \in P$, $D(p) \subseteq A$ and $D(F) = A \sim D(M)$ and $D(O) = A \sim D(Y)\}$

(3)(a) For every $p \in P, i \in I$, and every $\langle A, D\rangle \in M$,

$H(p^\frown i, \langle A, D\rangle)$ iff $D(i) \in D(p)$.

For every $s_1, s_2 \in S, m \in M$,

$H(\neg s_1, m)$ iff not $H(s_1, m)$;

$H(s_1 \,\&\, s_2, m)$ iff $H(s_1, m)$ and $H(s_2, m)$;

$H(s_1 \vee s_2, m)$ iff $H(s_1, m)$ or $H(s_2, m)$;

$H(s_1 \supset s_2, m)$ iff either not $H(s_1, m)$ or $H(s_2, m)$;

$H(s_1 \equiv s_2, m)$ iff either $H(s_1, m)$ and $H(s_2, m)$, or not $H(s_1, m)$ and not $H(s_2, m)$.

$\mathfrak{L}_4$ is the semantically structuralizable language with a semantically structured structuralization having this S, M, and H.

Because of the construction of the model set, $\models_{\mathfrak{L}_4} Ma \equiv \neg Fa$. Thus it is not just factually but logically true in this language that males are not females. A descriptively equivalent analysis of $\mathfrak{L}_4$ could in fact have been formulated by taking sentences '$[Ma \equiv \neg Fa]$', '$[Mb \equiv \neg Fb]$', '$[Mc \equiv \neg Fc]$', '$[Ya \equiv \neg Oa]$', '$[Yb \equiv \neg Ob]$', and '$[Yc \equiv \neg Oc]$' as what Kemeny and others have called 'meaning postulates'. After the specification of S the description would proceed by defining first a set M_0 of 'semi-models', the definition of M_0 being like (2)(a) with the conditions $D(F) = A \sim D(M)$ and $D(O) = A \sim D(Y)$ omitted. The rules of satisfaction are next defined as in (3)(a) except with M_0 as model set. Finally the real model set M is defined as the set of all members of M_0 satisfying the meaning postulates. The meaning-postulate approach to semantic description has its advantages in some circumstances, but is of course possible only for languages of sufficient expressive power. Had there been no binary connectives in $\mathfrak{L}_4$, for instance, the approach would not have been a viable alternative for it. In this sense the method of straight structural description is more powerful than the 'meaning postulate' approach.

EXAMPLE 5.5. *A Classical Predicate Language*. Extending $\mathfrak{L}_4$ by the addition of a quantificational apparatus and some relation symbols yields a member of the familiar family of first-order predicate calculi. For definiteness let us consider a language $\mathfrak{L}_5$ with three zero-place predicate constants (proposition symbols) P_1^0, P_2^0, P_3^0; four one-place predicate constants P_1^1, P_2^1, P_3^1, and P_4^1; two two-place predicate (relational) constants P_1^2, P_2^2; a three-place predicate constant P_1^3; and three individual constants a, b, and c. (Classical predicate languages often have for every n an infinity of n-ary predicate symbols called 'predicate variables'. But when the interest is in languages which could be learned directly and used for factual communication the assumption of some finite vocabulary of individual and predicate constants seems natural.)

The vocabulary of $\mathfrak{L}_5$ consists of the ten predicate constants, the three individual constants, and 'x', '$\prime$', '&', '$\vee$', '$\supset$', '$\equiv$', '$\forall$', '$\exists$', '[', and ']'. An 'x' followed by any finite number of primes is a *variable*. A *term* is a variable or an individual constant. A predicate constant with terms following it in the number of its superscript is an *atomic sentence*. The sentence set S of $\mathfrak{L}_5$ is the smallest set of finite strings of vocabulary elements containing the atomic sentences and such that if s_1 and s_2 are in S and ν is a variable then $\neg s_1$, $[s_1 \,\&\, s_2]$, $[s_1 \vee s_2]$, $[s_1 \supset s_2]$, $[s_1 \equiv s_2]$, $\forall \nu s_1$ and $\exists \nu s_1$ are also in S.

A semantically structured structuralization of $\mathfrak{L}_5$ must now be described starting with a model set M. Let M be the set of all ordered pairs $\langle A, D\rangle$ where A is any non-empty set (the domain of discourse for the model) and D is a function (the denotation mapping) assigning to each propositional constant either 1 or 0 (i.e. a truth value), and to each predicate constant with superscript $n \geqslant 1$ a set of n-tuples in A, and to each individual constant a member of A.

The rules of satisfaction will be described by saying first what it means for a 'completion' of a model to satisfy a sentence. For any $s \in S$, a *completion* for s of a model $\langle A, D\rangle \in M$ is a pair $\langle A, D^*\rangle$ in which D^* is a function constructed by extending the domain of D to include the variables occurring in s; specifically D^* assigns to each such variable a member of A, but is otherwise just like D. If $\langle A, D^*\rangle$ is a completion for s of $\langle A, D\rangle$, what it means for $\langle A, D^*\rangle$ to satisfy s is defined inductively for all subsentences s' of s by the following clauses: (i) if s' is a propositional constant then $\langle A, D^*\rangle$ satisfies s' if and only if $D^*(s') = 1$; (ii) if s' is of form $\phi^n \tau_1 \ldots \tau_n$ where ϕ^n is an n-ary predicate constant with $n \geqslant 1$ and $\tau_1, \ldots, \tau_n$ are terms then $\langle A, D^*\rangle$ satisfies

s' if and only if $\langle D^*(\tau_1), \ldots, D^*(\tau_n)\rangle \in D^*(\phi^n)$; (iii) if s' is of form $[s_1 \,\&\, s_2]$ then $\langle A, D^*\rangle$ satisfies s' if and only if $\langle A, D^*\rangle$ satisfies both s_1 and s_2; (iv)-(vii) the satisfaction rules for $\vee$, $\supset$, $\equiv$, and $\neg$ are similar (cf. Example 5.4 line (3)); (viii) if s' is of form $\forall v s_1$ then $\langle A, D^*\rangle$ satisfies s' if and only if for every completion $\langle A, D^{*\prime}\rangle$ in which $D^{*\prime}$ differs from D^* at most in what it assigns to v, $\langle A, D^{*\prime}\rangle$ satisfies s_1; (ix) if s' has the form $\exists v s_1$ then $\langle A, D^*\rangle$ satisfies s' if and only if for some completion $\langle A, D^{*\prime}\rangle$ in which $D^{*\prime}$ differs from D^* at most in what it assigns to v, $\langle A, D^{*\prime}\rangle$ satisfies s_1. A model $\langle A, D\rangle \in M$ *satisfies* $s \in S$ if and only if for any completion $\langle A, D^*\rangle$ for s of $\langle A, D\rangle$, $\langle A, D^*\rangle$ satisfies s. This completes the description of the satisfaction relation H. $\mathfrak{L}_5$ is the language with a semantically structured structuralization with sentence set S, model set M, and satisfaction relation H. The formal symbolic description will be dispensed with.

A couple of technical points. In conventional presentations of predicate logic a well-formed formula is often called a sentence only if it contains no free variables. In $\mathfrak{L}_5$ all well-formed formulae are sentences, a sentence with free variables being interpreted as though it were universally quantified over the entire sentence for all its free variables. This convention has been called the 'generality interpretation' (Kleene 1967, pp. 103ff.). The other point concerns the domains of discourse, which in the structuralization chosen for the above description were allowed to be any sets whatever. For the sake of structural economy the specification of M could if desired stipulate that all domains of discourse be sets of natural numbers. It follows from the well-known theorem of Lowenheim and Skolem that this requirement would not change the behavior of the language automaton.

In $\mathfrak{L}_5$ there happen to be no logical interdependencies among the constants. Variant languages with dependencies are obtainable by restricting M suitably. However, $\mathfrak{L}_5$ is expressive enough to allow the introduction of most such dependencies as meaning postulates. For instance, if P_1^1 and P_2^1 happened to mean maleness and femaleness like the M and F of Example 5.4, $\forall x\,[P_1^1 \equiv \neg P_2^1]$ could be taken as a meaning postulate. If P_1^0, P_2^0, and P_3^0 were like the P, Q, and R of Example 5.3, the meaning postulate $[\neg[P_1^0 \,\&\, P_2^0] \,\&\, [P_3^0 \supset [P_1^0 \vee P_2^0]]]$ would be called for. If P_1^2 had a meaning such as 'is older than', a transitivity postulate would be called for; and so on.

EXAMPLE 5.6. *A Modal Language*. Semantical rules are now known for a number of modal languages which attempt to formalize such concepts as necessity, possibility, and strict implication. As an example a language $\mathfrak{L}_6$ will be described which is essentially a pragmatic description of Lewis and

Langford's system $S5$ (1932) extended to quantification theory using semantical rules stated by Kripke (1959a, b). The sentence set S of $\mathfrak{L}_6$ is that of $\mathfrak{L}_5$ with the addition of the one-place sentence connective '$\Box$' (read 'it is necessary that'). The model set M is the set of all pairs $\langle G, K\rangle$ where K is a non-empty set of pairs $\langle A, D\rangle$ of the form of the models for $\mathfrak{L}_5$, A being the same in all pairs in K and G being a member of K. (Intuitively G is a possible state of affairs and K tells what would be logically possible if G were actual.) Satisfaction by a model $\langle G, K\rangle$ is defined by defining it first for the members of K. For any completion $\langle A, D^*\rangle$ for s of a member k of K, $\langle A, D^*\rangle$ satisfies a subsentence s' of s or not in accordance with the inductive clauses: (i)-(ix) (same as in Example 5.5); (x) if s' is of form $\Box s_1$ then $\langle A, D^*\rangle$ satisfies s' if and only if for every completion $\langle A, D^{*\prime}\rangle$ of every member k' of K such that $D^{*\prime}$ agrees with D^* in what it assigns to the variables of s, $\langle A, D^{*\prime}\rangle$ satisfies s_1. Finally, a model $\langle G, K\rangle$ satisfies s if and only if G satisfies s.

By virtue of the translatability of the Heyting intuitionist calculus into a form of modal logic, semantical rules are available for it too (Kripke 1965). Thus most of the better-known formal logician's languages can be so described as to fall within the semantically structuralizable language family.

5.3 A FRAGMENT OF ENGLISH

Semantical rules for fragments of English have been formulated by P. Suppes (1971) and his associates, by R. Montague (1974), and by others who have extended Montague's work in various directions (Partee, 1976). These semantical rules determine semantically structured language automata, and so from the present perspective may be regarded as logico-linguistic hypotheses about English. However, they were not necessarily originally conceived as such, and in the absence of clear rules of evidence for them their empirical validity is problematic.

An examination of evidence for and against Montague's language descriptions carried out under the logico-linguistic rules of evidence set forth in this book has shown them to be in conflict with observation much of the time (Cooper, in press). As hypotheses about English they must therefore be rejected in large part. But their descriptive inaccuracies do not diminish their historic importance, and they remain interesting examples of some of the descriptive techniques possible under the model-theoretic approach.

In illustration of this kind of semantics we present next a formal structural description of a small fragment of English. '$\mathfrak{L}_7$', as the fragment will be called, resembles some of the more accurately described portions of Montague's fragments, though our treatment of it will differ from Montague's in many respects. It contains the Detectives' Language $\mathfrak{L}_0$, with a change to the present tense, as a sublanguage.

Since the problem of ambiguity has not yet been dealt with, we will circumvent it for the time being by describing not ordinary English but 'disambiguated English'. This consists of English sentences with special ambiguity-breaking symbols inserted. Although in general such symbols could be labelled bracketings or any other disambiguating insertions, for the present fragment it will suffice to attach numeric superscripts to quantifiers and to *not* in such a way as to indicate their relative 'scope'. Thus if the ambiguous sentence

Every man loves some woman

is understood to mean that for every man there exists a woman whom he loves, the *every* has the larger scope and the disambiguation of it would be written

$\textit{Every}^1$ *man loves* $\textit{some}^0$ *woman*.

If on the other hand the intended interpretation is that some woman is such that every man loves her, the analyzed sentence would be

$\textit{Every}^0$ *man loves* $\textit{some}^1$ *woman*.

EXAMPLE 5.7. *A Fragment of English*

A. SYNTAX OF $\mathfrak{L}_7$

The task of the syntactic rules is to describe the set S of all (disambiguated) sentences of the fragment. The first few rules specify some vocabulary classes:

A1 (Proper Nouns).

$$S_1 = \{\textit{Coe}, \textit{Doe}, \textit{John}, \textit{Jane}, \textit{New York}\};$$

A2 (Countable Nouns)

$$S_2 = \{\textit{man}, \textit{woman}, \textit{widower}, \textit{widow}, \textit{horse}, \textit{frog}, \textit{car}, \textit{vehicle}\};$$

A3 (Predicate Adjectives)

$$S_3 = \{\textit{in}, \textit{out}, \textit{reliable}, \textit{beautiful}, \textit{married}\};$$

A4 (Intransitive Verbs, Singular Inflection)

$$S_4 = \{\textit{breathes}, \textit{reasons}, \textit{hibernates}\};$$

A5 (Transitive Verbs, Singular Inflection)

$$S_5 = \{\textit{loves}, \textit{sees}, \textit{hits}\};$$

A6 (Copula)

$$S_6 = \{\textit{is}\}.$$

A further rule establishes a 'proto-vocabulary' consisting of a denumerably infinite number of 'proto-nouns'. The proto-nouns will be endowed with properties something like those of the individual constants of the familiar formal languages. They will not appear in completed sentences, but will prove useful in the intermediate stages of sentence generation and semantic interpretation.

A7 (Proto-Nouns)

$$S_7 = \{c_1, c_2, c_3, \ldots\},$$

Next, some rules are needed to indicate how the foregoing vocabulary items mav be strung together to make longer structures. Letting '+' denote the interconcatenation operation (p. 48) and '∪' set-theoretic union, they may be written:

A8 (Transitive Proto-Verb Phrases)

$$S_8 = S_5 + S_7$$

Example: loves c_2;

A9 (Copulative Proto-Verb Phrases)

$$S_9 = S_6 + (S_3 \cup S_7)$$

Examples: is reliable, is c_2;

A10 (Basic Proto-Sentences)

$$S_{10} = S_7 + (S_4 \cup S_8 \cup S_9)$$

Examples: c_2 *hibernates,* c_2 *loves* c_{19}, c_3 *is reliable.*

With the basic proto-sentences so defined, the next step is to specify a larger set S_0 of 'proto-sentences' inductively with the basic proto-sentences serving as the basis of the induction. Let S_0 be the smallest set such that (i) $S_{10} \subseteq S_0$; and (ii) for every $s_1 \in S_1$, $s_2 \in S_2$, and $s_7 \in S_7$, and for all (possibly null) strings α and β, n being the number of superscripts occurring in $\alpha \frown \beta$,

A11 (Proper Noun Substitution)

If $\alpha^\frown s_7{}^\frown\beta \in S_0$ then $\alpha^\frown s_1{}^\frown\beta \in S_0$

Example: Since c_9 *loves* c_2 is a proto-sentence so is c_9 *loves Doe*;

A12 (Quantification by *every*)

If $\alpha^\frown s_7{}^\frown\beta \in S_0$ then $\alpha^\frown \mathit{every}^{n\frown} s_2{}^\frown\beta \in S_0$

Example: Since c_9 *loves* c_2 is a proto-sentence so is *Every*^{0} *man loves* c_2;

A13 (Quantification by *some*)

If $\alpha^\frown s_7{}^\frown\beta \in S_0$ then $\alpha^\frown \mathit{some}^{n\frown} s_2{}^\frown\beta \in S_0$

Example: Since *Every*^{0} *man loves* c_2 is a proto-sentence so is *Every*^{0} *man loves some*^{1} *woman*;

A14 (Quantification by *no*)

If $\alpha^\frown s_7{}^\frown\beta \in S_0$ and neither α nor β contains *not* or *no*, then $\alpha^\frown \mathit{no}^{n\frown} s_2{}^\frown\beta \in S_0$

Example: Since c_{12} *hibernates* is a proto-sentence so is *No*^{0} *horse hibernates*;

A15 (Negation)

If neither α nor β contains *not* or *no*, then

(a) if $\alpha^\frown \mathit{is}^\frown\beta \in S_0$ then $\alpha^\frown \mathit{is}^\frown \mathit{not}^{n\frown}\beta \in S_0$, and

(b) If $\alpha^\frown \nu^\frown\beta \in S_0$ where $\nu \in S_4 \cup S_5$ and ν' is the result of removing the singular inflection from ν, then $\alpha^\frown \mathit{does}^\frown \mathit{not}^{n\frown} \nu'^\frown\beta \in S_0$.

Example: Since *Every*^{1} *widow is some*^{0} *woman* is a proto-sentence so is *Every*^{1} *widow is not*^{2} *some*^{0} *woman*.

The sentence set of our fragment consists of all proto-sentences all of whose proto-nouns have been substituted for, i.e.

A16 (Sentences)

$$S = \{s_0 \mid s_0 \in S_0 \text{ and no } s_7 \in S_7 \text{ occurs in } s_0\}.$$

B. SEMANTICS OF $\mathfrak{L}_7$

The first semantical task is to describe a model set M for the fragment. All models in M will be given the same countably infinite domain of discourse A;

and since they all have the same domain it need not be exhibited explicitly in any of them. For all models, A will be understood to be the set of negative integers, i.e. $A = \{-1, -2, -3, \ldots\}$. This choice leaves 1 and 0 free to represent Truth and Falsity. M is now definable as the set of all 5-tuples $\langle D_1, \ldots, D_5 \rangle$ such that D_1 is a function with domain $S_1, \ldots, D_5$ is a function with domain S_5, and for every $s_1 \in S_1, \ldots, s_5 \in S_5$,

B1 (Denotation of Proper Nouns)

$$D_1(s_1) \in A$$

Example: In certain models *Coe* denotes -1, in certain others -2, etc.;

B2 (Denotation of Countable Nouns)

$$D_2(s_2) \subseteq A$$

Example: In certain models *frog* denotes $\{-3, -4, -9\}$;

B3 (Denotation of Predicate Adjectives)

$$D_3(s_3) \subseteq A$$

Example: In certain models *reliable* denotes the set of even negative integers;

B4 (Denotation of Intransitive Verbs)

$$D_4(s_4) \in \{0, 1\}^A$$

Example: In every model *hibernates* denotes some assignment of truth values to the negative integers;

B5 (Denotation of Transitive Verbs)

$$D_5(s_5) \in (\{0, 1\}^A)^A$$

Example: In every model *loves* denotes an assignment to the negative integers of assignments of truth values to the negative integers.

This completes the specification of M, except possibly for the insertion in B1-B5 of such assumptions as $D_2(\textit{widow}) \subseteq D_2(\textit{woman})$, $D_2(\text{widow}) \cap D_2(\text{widower}) = 0$, $D_3(\textit{in}) \cap D_3(\textit{out}) = 0$, etc. An alternative treatment would involve adding such assumptions later as meaning postulates.

The next semantical task is to describe the satisfaction relation H. H will be specified in terms of a 'denotation function' D assigning to each sentence a truth value relative to each model. D will in turn be specified with the help of a series of preliminary denotation functions. Let $\mathscr{P}A$ be the set of all subsets of A. Now define $D_6, \ldots, D_{10}$ to be functions with respective domains $S_6 \times M, \ldots, S_{10} \times M$ such that for every $m = \langle D_1, \ldots, D_5 \rangle \in M$,

B6 (Denotation of Copula)

$D_6(is, m)$ = the (unique) member of $(\{0, 1\}^A)^{A \cup \mathscr{P}A}$ such that (i) for every $a_1, a_2 \in A, D_6(is, m)(a_1)(a_2) = 1$ iff $a_1 = a_2$, and (ii) for every $A' \subseteq A$ and every $a \in A$, $D_6(is, m)(A')(a) = 1$ iff $a \in A'$

Example: In every model the denotation of *is* maps -4 to the truth value assignment in A in which -4 and only -4 is assigned Truth, and it maps $\{-3, -4, -7\}$ to the truth value assignment in which only $-3, -4$, and -7 are assigned Truth;

B7 (Denotation of Proto-Nouns)

For every $s_7 \in S_7, D_7(s_7, m) = -i$, where i is the value of the subscript on s_7

Example: In every model c_1 denotes -1, c_2 denotes -2, etc.,

B8 (Denotation of Transitive Proto-Verb Phrases)

For every $s_5 \in S_5, s_7 \in S_7$,

$$D_8(s_5 \frown s_7, m) = D_5(s_5)(D_7(s_7))$$

Example: In every model *loves* c_2 denotes the truth-value assignment to the negative integers obtained by applying the denotation of *loves* to -2;

B9 (Denotation of Copulative Proto-Verb Phrases)

For every $s_i \in S_i, i = 3, 7$,

$$D_9(is \frown s_i, m) = D_6(is, m)(D_i(s_i, m))$$

Example: In every model *is reliable* denotes the truth-value assignment in A obtained by applying the denotation of *is* to the denotation of *reliable*;

B10 (Truth Value of Basic Proto-Sentences)

For every $s_7 \in S_7$ and every $s_i \in S_i, i = 4, 8, 9$,

$$D_{10}(s_7 \frown s_i, m) = D_i(s_i, m)(D_7(s_7, m))$$

Example: c_5 *reasons* denotes Truth in those models in which the denotation of *reasons* maps — 4 to 1.

Next D_{10} will be extended by an inductive definition to a function D_0 assigning truth values to proto-sentences for specified models. Let D_0 be the (unique) function from $S_0 \times M$ into $\{0, 1\}$ such that (i) D_0 agrees with D_{10} on $S_{10} \times M$, and (ii) for every $m = \langle D_1, \ldots, D_5 \rangle \in M$, for every $s_1 \in S_1$, $s_2 \in S_2$, and $s_7 \in S_7$, and for all (possibly null) strings α and β, if $\alpha \frown s_7 \frown \beta \in S_0$ and n is the number of superscripts occurring in $\alpha \frown s_7 \frown \beta$ then:

B11 (Truth Values of Proto-Sentences containing Proper Nouns)

If $D_1(s_1) = D_7(s_7)$ then

$$D_0(\alpha \frown s_1 \frown \beta, m) = D_0(\alpha \frown s_7 \frown \beta, m)$$

Example: In any model, if c_9 *loves* c_2 is true and *Doe* denotes — 2 then c_9 *loves Doe* is true;

B12 (Truth Values of Proto-Sentences containing *every*)

$D_0(\alpha \frown every^n \frown s_2 \frown \beta, m) = 1$ iff for every $s_7 \in S_7$ such that $D_7(s_7) \in D_2(s_2)$, $D_0(\alpha \frown s_7 \frown \beta, m) = 1$

Example: *Every*^{0} *man loves* c_2 holds true in models such that, for every i such that $-i$ is in the denotation of *man*, c_i *loves* c_2 is true;

B13 (Truth Values of Proto-Sentences containing *some*)

$D_0(\alpha \frown some^n \frown s_2 \frown \beta, m) = 1$ iff there exists an $s_7 \in S_7$ such that $D_7(s_7) \in D_2(s_2)$ and $D_0(\alpha \frown s_7 \frown \beta, m) = 1$

Example: *Every*^{0} *man loves some*^{1} *woman* is true in models such that, for some i such that $-i$ is in the denotation of *woman*, *Every*^{0} *man loves* c_i is true;

B14 (Truth Values of Proto-Sentences containing *no*)

$D_0(\alpha \frown no^n \frown s_2 \frown \beta, m) = 1$ iff there does not exist an $s_7 \in S_7$ such that $D_7(s_7) \in D_2(s_2)$ and $D_0(\alpha \frown s_7 \frown \beta, m) = 1$

Example: *No*^{0} *horse hibernates* is true in all models in which c_i *hibernates* is false for all i such that $-i$ is in the denotation of *horse*;

B15 (Truth Values of Negated Proto-Sentences)

(a) $D_0(\alpha^\frown is^\frown not^{n\frown}\beta, m) = 1 - D_0(\alpha^\frown is^\frown\beta, m)$

(b) for every $\nu \in S_4 \cup S_5$, if ν' is the result of removing the singular inflection from ν then $D_0(\alpha^\frown does\ not^{n\frown}\nu'^\frown\beta, m) = 1 - D_0(\alpha^\frown\nu^\frown\beta, m)$

Example: *John does not*^{0} *reason* is true in just those models in which *John reasons* is false.

It remains only to restrict attention to truth values of sentences as opposed to all proto-sentences.

B16 (Truth Values of Sentences)

D is the restriction of D_0 to $S \times M$.

The denotation function D is in effect the wanted satisfaction relation H; however, H can if desired be defined explicitly by the postulate

B17 (The Satisfaction Relation)

For every $s \in S, m \in M$,

$H(s, m)$ iff $D(s, m) = 1$.

With S, M, and H so specified, a semantically structured language automaton may be constructed along the lines of Definition 4.1 (i). $\mathfrak{L}_7$ is that automaton's equivalence class. This completes the formal description of the fragment.

The derivational history of any sentence in the fragment can be diagrammed after this fashion:

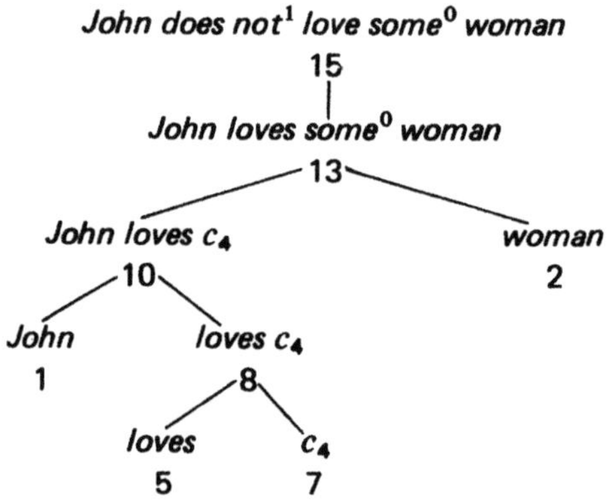

Actually the diagram has two interpretations. If the numbers at the nodes are taken to refer to the syntactic rules A15, A13, A10, etc., then the tree shows a possible syntactic derivation of the sentence – a way of applying the syntactic rules to prove that the sentence is indeed in S. If on the other hand the numbers are understood to refer to the semantic rules B15, B13, etc., the tree indicates the steps needed to deduce the sentence's truth value in any given model. Such diagrams are heuristic devices only, however. They play no essential role in logico-linguistic theory proper.

The semantical rules of our description are too 'definite' in that they sometimes assign truth values under conditions in which there really are no clear truth values. For instance, it is not clear (at least in my dialect) what the truth value of *Every*^{0} *frog hibernates* would be in a world with no frogs. This descriptive shortcoming could be repaired by allowing the satisfaction relation to remain undefined on occasion, with concommitant incompleteness allowed elsewhere in the description. We will forego that nicety here, however, postponing discussion of the role of incompleteness to § 9.4.

To verify the empirical adequacy of the description, the first step would be to test the syntactic rules A1-A16. That standard exercise will not be carried out here, but would typically consist of 'generating' strings in S with the help of A1-A16 and submitting the results to English speakers to obtain judgements on their well-formedness. Such testing is not without its problems, one of which is the thorny matter of what to do about strings like *New York loves Jane* and other anomalous expressions, but that issue turns out to be somewhat related to the frogless world problem and will be discussed later. The second step would be to test the semantic rules *via* their pragmatic consequences. This step will not be carried out here systematically either, but a couple of examples will illustrate what is involved.

A typical check on B1-B16 would involve the examination of some logical relationship among specified sentences of S. Consider

No^0 *frog reasons*

John reasons

John is no^0 *frog* .

Using ourselves as informants we note that it is hard to imagine any rational English speaker, no matter what his information state, believing the sentences above the line without also believing the one below. In other words, informant technique shows the implication to be valid in the sense that the basic evidence property involved is a property of English. This completes the observation; what must be determined next is whether the formal language description predicts it. Now, from rules B1-B17 it may be deduced mathematically that *John is no⁰ frog* must hold true in any model in which *No⁰ frog reasons* and *John reasons* hold true. From this it can be deduced further that for every $z \in Z$, if B(z, *No⁰ frog reasons*) and B(z, *John reasons*) then B(z, *John is no⁰ frog*). But this is the basic evidence property English was observed to possess. The example is therefore a confirming instance of the description.

As a second example consider the sentences

Every^1 *man loves some*^{0} *woman*

No^0 *man loves Mary*

* ———————————

These sentences are logically compatible on the pragmatic level, for there is no difficulty imagining a rational English speaker believing them both simultaneously. That is the observation. But it can also be shown formally by B1-B17 that they are compatible. Hence the example constitutes another bit of confirming evidence.

Further testing would consist of a similar scrutiny of many more such examples, chosen either at random or according to some systematic sampling plan. Any logical relationship predicted to exist by the formal rules of the description but not observed to exist, or predicted not to exist but observed to exist, would constitute counterevidence to the description. I have examined a moderate number of examples without finding any clear counterevidence, encouraging the hope that the description may be approximately accurate. But as is the case with any scientific hypothesis, there is no guarantee that further testing would not turn up some counterevidence calling for revision or rejection of the rules.

As for the role of philosophical disputation in logico-linguistic analysis, suppose a critic were to object to the foregoing description on the grounds that it gives every possible world the same denumerably infinite domain of discourse. The critic might maintain that by his metaphysical lights different possible worlds have different universes of discourse of very different sizes. We would accept this as a valid criticism of the description only if he were able to back up his metaphysics with some actual counterevidence in the form of a conflict between predicted and observed basic evidence properties brought about by the assumption of a common domain of discourse. If the critic were unable to produce such counterevidence, we would suspect that he is just talking about a possible alternative structuralization in the same equivalence class as ours. If so there is no formidable metaphysical point at issue, but only a matter of descriptive taste.

5.4 Semantics and Deductive Logic

The traditional semantic account of logical implication is that s_1 logically implies s_2 if and only if s_2 would hold true in any circumstances (possible worlds) in which s_1 held true. Which is better, this or the pragmatic explanation of implication in terms of information states and believability? Fortunately there is no conflict, for semantically structuralizable languages at any rate. In these languages, which are probably the only languages for which the semantic criterion makes sense, it can be shown that semantic and pragmatic implication come to the same thing. Thus the philosophies behind the two approaches are mutually reinforcing, though the pragmatic is more broadly applicable than the semantic.

DEFINITION 5.2. S_1 *semantically implies* s in a semantically structuralizable language $\mathfrak{L}$ (abbreviated $S_1 \models_{\mathfrak{L}} s$) if an only if for some (every) semantically structured structuralization $\langle S, Z, L, B\rangle \in \mathfrak{L}$ with satisfaction relation H, $S_1 \subseteq S$ and $s \in S$ and

$$\bigcap_{s_1 \in S_1} H[s_1] \subseteq H[s].$$

DEFINITION 5.3. S_1 is *semantically inconsistent* in a semantically structuralizable language $\mathfrak{L}$ ('$\not\models_{\mathfrak{L}} S_1$') if an only if for some (every) semantically structured structuralization $\langle S, Z, L, B\rangle \in \mathfrak{L}$ with satisfaction relation H,

$$\bigcap_{s_1 \in S_1} H[s_1] = 0.$$

THEOREM 5.1. ('Pragmatic Completeness Theorem'). Let $\mathfrak{L}$ be a semantically structuralizable language with sentence set S. Then for every $S_1 \subseteq S$ and every $s \in S$,

(i) $S_1 \models\!\!\equiv_{\mathfrak{L}} s$ if and only if $S_1 \models_{\mathfrak{L}} s$;

(ii) $\not\models\!\!\equiv_{\mathfrak{L}} S_1$ if and only if $\not\models_{\mathfrak{L}} S_1$.

THEOREM 5.2. If $\mathfrak{L}$ is a semantically structuralizable language then $\mathfrak{L}$ is stable, receptive, and intransigent.

COROLLARY. Let $\mathfrak{L}$ be a semantically structuralizable language with sentence set S, let $s \in S$, let $\bar{s}$ be any finite sequence of sentences in S, and let S_1 be the set of all sentences occurring in $\bar{s}$. Then

(i) $\bar{s} \overrightarrow{\models}_{\mathfrak{L}} s$ if and only if $S_1 \models_{\mathfrak{L}} s$;

(ii) $\bar{s} \overrightarrow{\not\models}_{\mathfrak{L}} s$ if and only if $\not\models_{\mathfrak{L}} S_1 \cup \{s\}$.

It is an interesing speculation that the reason why the differences between the logical and quasi-logical relationships have remained largely unnoticed in traditional logic is that the commonly studied formal languages all happen to belong to the semantically structuralizable language family within which there are no such differences.

There are other ways of characterizing the logical relationships in semantically structuralizable languages. Recall that the model set M of a semantically structured language automaton is itself a member of the state set. Let us refer to M as the *nescient state* of the automaton and any member $z \in Z$ that has no proper subsets in Z an *omniscient state* of the automaton. Intuitively the nescient state is a state of total ignorance, while a language user in an omniscient state would already know so much he could not learn anything new.

THEOREM 5.3. Let $\mathfrak{L}$ be a semantically structuralizable language and let $\langle S, Z, L, B \rangle$ be a semantically structured structuralization of $\mathfrak{L}$. Then for any $S' \subseteq S$ and any s in S,

(i) $S' \models_{\mathfrak{L}} s$ if and only if: for every omniscient z in Z, if $B(z, s')$ for all s' in S' then $B(z, s)$;

(ii) $\not\models_{\mathfrak{L}} S'$ if and only if there exists no omniscient z in Z such that $B(z, s')$ for all s' in S'.

Proof: The 'only if' part of (i) is immediate from the definition of logical implication. To prove the contrapositive of the 'if' part of (i), suppose that S' does not logically imply s in $\mathfrak{L}$. Then by Theorem 5.1, S' does not semantically imply s in $\mathfrak{L}$. Thus there exists an m in $\bigcap_{s' \in S'} H[s']$ that is not in $H[s]$. For this m, $\{m\}$ is omniscient in Z, $B(\{m\}, s')$ for all s' in S', and not $B(\{m\}, s')$. This contradicts the conclusion. The proof of (ii) is similar.

THEOREM 5.4. Let $\mathfrak{L}$ be a semantically structuralizable language and let $\langle S, Z, L, B \rangle$ be a semantically structured structuralization of $\mathfrak{L}$ with model set M. Then for any finite sequence $\bar{s}$ of members of S and any s in S, $\bar{s} \overrightarrow{\models}_{\mathfrak{L}} s$ if and only if either $B(\bar{L}(M, \bar{s}), s)$ or $\bar{L}$ is undefined for M and $\bar{s}$.

Proof: The 'only if' is immediate. For the 'if' direction suppose first that L is undefined for M and $\bar{s}$. Upon inspecting the characterization of L in the definition of 'semantically structured' it becomes clear that if $\bar{L}$ is undefined for M and $\bar{s}$ it must also be undefined for any subset of M and $\bar{s}$, from which it follows that $\bar{s} \overset{\rightarrow}{\models}_{\mathfrak{L}} s$. Next suppose $\bar{L}$ is defined for M and $\bar{s}$ and $B(\bar{L}(M, \bar{s}), s)$. For any non-empty subset z of M, if $\bar{L}$ is defined for z and $\bar{s}$ it is clear from the characterization of L that $L(z, \bar{s}) \subseteq L(M, \bar{s})$. From this it again follows that $\bar{s} \models_{\mathfrak{L}} s$.

THEOREM 5.5. A semantically structuralizable language has an incomplete logic (and hence is informative) if and only if each of its semantically structured structuralizations has more than one model in its model set.

5.5 Axiomatic Language Descriptions

Since antiquity the axiomatic method has been a powerful force in logic, and the most familiar way of specifying a system of logic is probably still that of laying down axioms and rules of inference. In pragmatic language analysis the axiomatic approach provides an alternative way of formulating language descriptions that avoids the need for detailed semantical or other structural rules. The general idea is to select from the language to be described certain sentences as axioms, and to use these in conjunction with appropriate rules of inference to specify the logic of the language. The description is then completed with an indication of which language among those having that logic the language of interest is.

Of course even semantic language descriptions are 'axiomatic' in the sense that they are written in a set-theoretic language having its own axioms and rules of inference. However, that is not the meaning of 'axiomatic' intended here. What is meant is the use as axioms of sentences taken directly from the language to be described, and rules of inference operating directly among these sentences. If the metalanguage is formalized it may indeed have its own axioms and rules of inference. The metalinguistic apparatus may or may not be used to specify axioms and rules of inference in the object language; when it is the result is an 'axiomatic language description' in our sense.

Axiomatic language descriptions are quasi-behavioral. They consist largely of a statement of the language's (behavioral) logical relationships described axiomatically without reference to internal language automaton structure. A structural element enters the description, if at all, only at the point where it is said which language with the specified logic is the one being described. Often this may be indicated simply by restricting attention to some predefined language family such as the family of stable and receptive languages or the family of all semantically structuralizable languages. However, it is one of the disadvantages of the axiomatic approach that it is not always immediately apparent that there exists one and only one language with the specified logic in the indicated family. Because of this an axiomatic language description must ordinarily be accompanied by some sort of auxiliary argument demonstrating existence and uniqueness.

There are semantically structuralizable languages which cannot be described axiomatically, and axiomatically describable languages which are not semantically structuralizable. For languages amenable to both semantic and axiomatic description, a proof of the descriptive equivalence of the two kinds of description is a pragmatic counterpart of the traditional semantic completeness proof.

After specifying the language's sentence set S, an axiomatic language description sets forth a list of axioms or axiom schemata telling which members of S are in the axiom set. Accompanying rules of inference are also specified. A derivability relation $\vdash$ for the language is then defined in the customary manner via the notion of a formal proof using the axioms and rules of inference. A refutability property $\Vdash$ is next specified, either as the null property in case there are no inconsistent sets of sentences in the language, or by using Post's criterion (§ 4.3) in case there are. The logic of the language to be described is then $\langle S, \vdash, \Vdash \rangle$.

A class of languages is next defined, or a predefined class named, and the language of interest declared to be a member of this class. If there is only one language in the class with the specified logic, the description is complete (except possibly for the auxiliary argument proving this). If on the other hand there are many, one in particular must be singled out to complete the description.

When the class of languages happens to be the semantically structuralizable language family, the following fact is often useful in specifying which member of the family the language of interest is. Let us call one language a ***state-sublanguage*** of another language if and only if some structuralization of the first is a Z-subautomaton of some structuralization of the second. A language is ***state-greatest*** in a class of languages if and only if every member of the class is a state-sublanguage of it, and ***state-least*** if and only if it is a state-sublanguage of every member. Though the proof is too lengthy to present, it can be shown rigorously that for any logic $\langle S, \vdash, \Vdash \rangle$ specified axiomatically after the fashion of the preceding paragraphs, the class of semantically structuralizable languages with that logic is non-empty and has a unique state-greatest member and a unique state-least member. This fact allows many semantically structuralizable languages to be described axiomatically simply by stating in closing that the language of interest is the state-greatest (or state-least) semantically structuralizable language with the specified logic.

EXAMPLE 5.8. *An Axiomatic Description of* $\mathfrak{L}_4$. A descriptively equivalent alternative description of the census-takers' language of Example 5.4 can be given as follows. The sentence set S is described as in Example 5.4. Taking the atomic sentences '*Ma*', '*Mb*', '*Mc*', '*Fa*', etc. as the propositional symbols, any standard axiom system for the classical propositional calculus is next stated; the thirteen axiom schemata and ***Modus Ponens*** rule of inference stated by Kleene (1967, p. 387) will do. To these are added the special (meaning-)-postulates $[Ma \equiv \neg Fa]$, $[Mb \equiv \neg Fb]$, $[Mc \equiv \neg Fc]$, $[Ya \equiv \neg Oa]$, $[Yb \equiv \neg Ob]$, and $[Yc \equiv \neg Oc]$ to complete the axiom system for $\mathfrak{L}_4$. Now, for any $S' \subseteq S$ and any $s \in S$ let $S' \vdash s$ if and only if s is derivable from S' and members of the axiom set by a finite number of applications of the rules of inference. Further let $\Vdash S'$ if and only if $S' \vdash s$ for every s in S. $\mathfrak{L}_4$ is the state-least semantically structuralizable language whose logic is $\langle S, \vdash, \Vdash \rangle$. This completes the description. It follows by an extension of the standard completeness proof for the propositional calculus that the $\mathfrak{L}_4$ so described is identical with the $\mathfrak{L}_4$ of Example 5.4.

5.6 OTHER LANGUAGE FAMILIES

Semantic and axiomatic language descriptions have been discussed here in some detail because of the prominence of semantics and axiomatics in current logical thinking. They are by no means the only conceivable approaches to logico-linguistic analysis, however. The possible approaches are limited only by the ingenuity of the language scientist in devising linguistic hypotheses.

One approach to the description of language automaton structure which is at least superficially different from the semantic approach involves specifying the rules of verbal behavior as algorithms instead of abstract set-theoretic entities. Under this kind of 'procedural semantics' a language automaton for a language would be specified by a set of rules having the character of a computer program written to make a computer behave like the appropriate language automaton. The rules could be formulated in any convenient command language, those that come to mind first being some of the more recent list-handling and other symbol-manipulation computer languages with strong recursive capabilities. The resulting program would be a legitimate structural language description, for it would specify a language automaton of which the language of interest is understood to be the equivalence class. This descriptive approach might entail some loss of representational power to the extent that it excludes learning and belief rules which are not computable functions. However, not everyone would agree that this is a real loss, and in any case there may be quasi-algorithmic methods that get around the restriction.

A related family of descriptive techniques is suggested by recent experimentation in artificial intelligence. In attempting to simulate human or near-human language handling abilities on a computer, a number of workers in artificial intelligence have been led to represent information in computer memory by systems of list structures, trees, and other kinds of address-pointer networks. The computer stores new information in the network by adding to it nodes, labels, or arrows, and when the necessity arises of seeing whether a particular fact is currently stored the computer finds out by address-pointer-chasing search techniques. All these ideas should lend themselves to algorithmic descriptions of language automata.

In representing these memory structures abstractly, the branch of mathematics usually called upon is graph theory – typically the theory of directed labelled graphs. Since graphs are perfectly definite mathematical entities, they suggest themselves as another kind of mathematical mechanism which the black box might be postulated to contain. Thus one might be led to investigate 'graph-structured language automata' and corresponding 'graph-structuralizable languages'. This language family seems worthy of study not only because of the generality of the graph concept but also because of the possibility of a fruitful interplay between logico-linguistic theory and computer experimentation.

There are other possibilities. One is the exploitation of what is known in automata theory about the fine structure of automata when represented as a logical network of simplified elementary automata. It is conceivable that the rules for some languages might be illuminated when broken down into their fine structure. Another is the adaptation of cognitive models proposed by psychologists which, though performance-oriented, might be at least suggestive of ways of describing logico-linguistic competence. More generally, the philosophy of structuralism which has recently pervaded parts of psychology and the life sciences as well as linguistics may eventually lead to the discovery of hypothetical constructs that constrain thought processes and hence may have a place in structural language descriptions. Which approach is best suited to the analysis of natural languages, if there is any single best approach, remains to be seen.

5.7 LOGIC AS A BRANCH OF LINGUISTICS

The philosophy of logicism which culminated in Russell and Whitehead's *Principia Mathematica* is sometimes summed up in the following aphorism:

LOGISTIC THESIS. *Mathematics is a branch of deductive logic.*

Though a shameless oversimplification, there are elements of truth in the slogan which could be spelled out more precisely.

A loosely analogous relationship can be asserted of logic and language studied in the pragmatic perspective:

LOGICO-LINGUISTIC THESIS. *Deductive logic is a branch of linguistics.*

This is again an oversimplification, but it does have some justification in these particulars: (1) logical implication, inconsistency, and the other logical and dynamic relationships are all definable in terms of linguistic primitives, notably information states and language automata (2) the model-theoretic development of logic is subsumed under the study of semantically structuralizable languages; and (3) the traditional axiomatics of logic finds its place in axiomatic language descriptions. That linguistics is not included in deductive logic is evident from the fact that every language has a unique logic but not *vice versa*.

It might be objected to the Logico-Linguistic Thesis that linguists have concerned themselves mainly with describing existing natural languages while mathematical logicians have invented their own formal languages. But this is historic accident. Some day logicians may become more involved with the logic of natural languages (this is already starting to happen) while linguists become more involved with exploring the theoretical properties of formalized approximations to the natural languages. There is evidence that linguists are already becoming more concerned with the logic of natural language. In any case there is nothing internally inconsistent in the view that logicians and linguists are, or should be, engaged in essentially the same business, though the logicians are a little more specialized.

5.8 SYNTAX, SEMANTICS, PRAGMATICS

Following Morris and Carnap, linguistic subject matter has traditionally been classified as either 'pragmatic', 'semantic', or 'syntactic'. According to Carnap these are fields of increasing abstraction and decreasing inclusiveness (Carnap 1942, p. 9):

> ... If in an investigation explicit reference is made to the speaker, or, to put it in more general terms, to the user of a language, then we assign it to the field of *pragmatics*. ... If we abstract from the user of the language and analyze only the expressions and their designata, we are in the field of *semantics*. And if, finally, we abstract from the designata also and analyze only the relations between the expressions, we are in (logical) *syntax*.

This conventional breakdown is not entirely satisfactory because it treats semantics as though it were a separable study, whereas we have argued that semantics has only a subordinate role as one possible kind of hypothesis about an essentially pragmatic entity. Semantics can never be a science in its own right because the only observational data by which semantic hypotheses can be tested is pragmatic; and as soon as one invokes the pragmatic theory needed to bring this data to bear on semantic questions, one immediately finds oneself on the more inclusive pragmatic level. Moreover, it is not even clear that all languages *have* a semantics in any widely accepted sense; certainly not all are semantically structuralizable under the present definition.

If a tripartite breakdown must be made, a better one would be this:

LEVEL I. Syntax.

LEVEL II. Logico-Linguistic Pragmatics
- A. Structural Analysis (e.g. semantics)
- B. Behavioral and Quasi-Behavioral Analysis (e.g. axiomatics)

LEVEL III. Higher Pragmatics

In this schematization the syntactic level is concerned exclusively with rules of formation, not logical relationships. There is no semantic level. The next level of inclusiveness that forms a separable subscience is that of logico-linguistic pragmatics, the level with which the present work is concerned. Language analysis on this level can be structural, behavioral, or in between (quasi-behavioral). Semantics fits in as *one* approach to structural analysis; axiomatics is *one* approach to quasi-behavioral analysis. Carnap's 'logical syntax' is best classed as part of what is needed for an axiomatic language description. The third and most inclusive level is higher pragmatics. It subsumes, in addition to logico-linguistic pragmatics, whatever intra-language universal rules of language there may be outside of the Language Automaton. It is possible that when this third level is better understood it will be found to separate into further sublevels.

CHAPTER 6

MEANING

The conviction that sentences have a 'meaning' or 'content' distinct from their physical structure is so firmly ingrained in the common sense view of language that any theory of linguistic communication failing to lend substance to this conviction is apt to appear incomplete and unsatisfying. It is a common point of view, in fact, that the first requirement to be made of a theory of language that ventures beyond syntax is a clarification of the nature of meaning.

The logico-linguistic theory presented so far in this work may seem rather far from the idea of meaning. It is nevertheless reinterpretable as a theory of meaning, when meaning is identified with information content at least. The burden of this chapter, which may be omitted without serious loss of continuity, is to demonstrate this.

6.1 PURPORTS AND IMPORTS

The concept of meaning has been an object of philosophic controversy since antiquity. It is so surrounded by unresolved debates and controversial distinctions that the modern investigator is sometimes well advised to dissociate himself from all precedent and define his terms afresh. If the explication of 'meaning' to be proposed conflicts with other explications of the word, or has something in common with prior explications of 'content', 'gloss', 'sense', 'signification', etc., that is neither here nor there. It is the explication that is important, not the label attached to it.

Attention will be restricted here to the meaning of whole sentences. Meaning as it applies to words, phrases, or other sentence parts may well be a derivative notion definable in terms of sentence meaning, but the extension of the theory into that area will not be explored here. Also, no attempt will be made (at first) to make the meaning representation economical. We seek first a representation of meaning which is inclusive, even if wasteful, in order to ensure that anything ordinarily thought of as part of the information content of a sentence is determinable from the representation, saving Ockham's razor for later.

Interestingly, the leading definition of meaning in Webster's *Third New International Dictionary* has two parts which we quote here in full:

meaning: 1a. the thing one intends to convey by act or esp. by language: PURPORT (do not mistake my ~);

1b. the thing that is conveyed or signified esp. by language: the sense in which something (as a statement) is understood: IMPORT (what is its ~ to you).

Although one is in no way constrained to follow Webster when framing a technical definition, this two-faceted view of meaning is nevertheless suggestive. It amounts to a split of the concept of meaning into the 'sender-meaning' of a message (purport) and its 'receiver-meaning' (import). Let us consider sender-meaning first.

To characterize a sentence's purport – its meaning with respect to its potential speaker – one must ask what property the speakers must have in order to be capable of affirming the sentence honestly. Clearly the essential property is that they be in possession of information on the basis of which they can know or believe the sentence. For if one knew how a speaker had to be informed to be able to assert the sentence, one would then know from the speaker-viewpoint what information it expressed – namely, the information held in common by all and only those in a position to assert it.

Since telling how a speaker would have to be informed to assert a sentence is to reveal the (speaker-)content of the sentence, the purport of a sentence is fully specified as soon as the property of being so informed is fully described. In other words, the purport of a sentence is a property of information states – the property of supporting belief in the sentence. It is the property the speaker purports to have by uttering the sentence, and expects his hearer to deduce that he has.

Using the standard set-theoretic representation of a property as the set of all entities possessing the property, the concept of a purport may be defined as follows.

DEFINITION 6.1. Let $\mathfrak{A} = \langle S, Z, L, B \rangle$ be a language automaton and let $s \in S$. Then the *purport* p of s in $\mathfrak{A}$ is

$$\{z \in Z \mid B(z, s)\}.$$

(If $\mathfrak{A}$ is not complete the purport p of s may be represented by a partial function $p: Z \to \{1, 0\}$ where $p(z) = 1$ if B is defined for z and s and $B(z, s)$,

$p(z) = 0$ if B is defined for z and s and not $B(z, s)$ and $p(z)$ is undefined if B is not defined for z and s.)

Turning to imports, it seems obvious that the receiver-meaning of a sentence is evidenced in the state changes it brings about or could bring about in those who hear or read it. Thus the import of a sentence is seen only in its effects on the states of real or potential receivers. If one could say, for every conceivable fund of information a hearer might possess, how his information would be affected by hearing and accepting a certain message, one would know what the informative (receiver-)content of the message was. One is led to represent the import of a sentence as a transformation of information states – mathematically a mapping from the state set into itself. This equates the receiver-meaning of a sentence with a complete account of what accepting the sentence would do to any conceivable hearer.

DEFINITION 6.2. Let $\mathfrak{A} = \langle S, Z, L, B \rangle$ be a language automaton and let $s \in S$. Then the *import* of s in $\mathfrak{A}$ is the function i: $Z \to Z$ such that $i(z) = L(z, s)$ for all $z \in Z$.

(For incomplete $\mathfrak{A}$ an import is a partial function.) Note that a sentence's import is not the state change it would occasion in a particular language user at a particular time. This misreading of the definition would make a sentence meaningless to anyone who happened to believe it already! The import is the mapping as a whole.

To take into account both sender- and receiver-meaning, the meaning of a sentence may now be described as all the characteristics of the sentence determined by both its purport and import.

DEFINITION 6.3. Let $\mathfrak{A} = \langle S, Z, L, B \rangle$ be a language automaton and let $s \in S$. Then the *meaning* of s in $\mathfrak{A}$ is the pair $\langle p, i \rangle$ where p and i are respectively the purport and import of s in $\mathfrak{A}$.

Ideas related to this way of representing meaning are to be found in the literature. MacKay distinguishes meaning to the sender from meaning to the receiver, describing meaning to the receiver of a message as "its selective function on the range of the recipient's states of conditional readiness for goal-directed activity" (1969, e.g. pp. 24, 71, 84). The import idea is also implicit in the literature on the semantics of programming languages and has been stated clearly by Scott and Strachey (1971). These authors do not introduce any notion comparable to our 'purport', but this is not necessarily a significant omission so far as programming languages are concerned. Indeed, the inapplicability of the purport concept to computer languages characterizes rather well the difference between command languages and languages usable for the two-way communication of factual information.

6.2 PURPORT–IMPORT GLOSSARIES

To know the meanings of the sentences of a language is to know the language in an important sense. This suggests that it might be useful to have a way of representing in a single abstract entity the sentences of a language together with the meaning of each sentence. Since a listing of messages with their meanings in document form is a 'glossary' ('dictionary', etc.), it is natural in the abstract development to look for a mathematical construct akin to a glossary for whole sentences.

When meanings are represented by purport–import pairs, a glossary is in the abstract an assignment of such a pair to each sentence of the language. So long as all the sentences are clear and unambiguous one and only one pair will be assigned to each sentence. The result is a 'purport–import glossary'.

DEFINITION 6.4. The *purport–import glossary* of a language automaton $\mathcal{A} = \langle S, Z, L, B \rangle$ is the function $\mathcal{G}$ with domain S such that for any $s \in S$, $\mathcal{G}(s)$ is the meaning of s in $\mathcal{A}$.

It is obvious that for every language automaton there exists one and only one purport–import glossary. It can be shown too that it is impossible for two distinct language automata to have the same glossary. This establishes that the correspondences between language automata and purport–import glossaries is one–one: when meanings are represented by purport–import pairs *the theory of meaning is in a sense isomorphic to the theory of language automata*. What can be said behaviorally about a language can in principle always be restated meaning-theoretically and *vice versa*.

It would have been feasible, as an alternative to the approach taken in earlier chapters, to develop the entire theory of language and logic as a theory of meaning. The primitives of the theory of meaning would have been, after information states, purport and imports. A language would be defined to be an equivalence class of purport–import glossaries. A structural language description would contain detailed rules projecting for each sentence a structure representing the purport and the import of the sentence. Logical and quasi-logical relationships would be defined to hold among sentence meanings – the logical relationships would be relationships among purports, the quasi-logical would also involve imports.

The significance of this alternative development as a theory of meaning is that logico-linguistic theory is not really so exclusively 'behavioral' in flavor as its presentation in this work might suggest. Those who do not find behaviorism to their taste as an approach to language analysis might well prefer the meaning-theoretic development of the theory, though of course in view of the isomorphism the difference is one of emphasis and interpretation rather than substance.

6.3 SPECIALIZED GLOSSARIES

Purport–import glossaries are *comprehensive* in the sense that they tell everything there is to know about a language on the logico-linguistic level – one could always reconstruct an entire language automaton from a knowledge of its purport–import glossary alone. Other kinds of glossaries which tell less about sentence meanings, for example glossaries specifying purports alone or imports alone, are *specialized* in the sense that they describe only certain aspects of meaning. A specialized glossary would not in general be a sufficient basis for reconstructing the associated language automaton in its entirety, or at least it would not be sufficient without some supplementary data about the nature of the language in question. Specialized glossaries may be useful nonetheless, either because they describe an aspect of meaning that happens to be of particular interest or because the needed supplementary data about the language is available which would allow it to be reconstructed from the glossary.

Suppose for example a language theorist were willing to confine his attention to stable and receptive languages. This is not an implausible supposition; some might even argue that stability and receptiveness are such compelling simplifying assumptions that they should be made part of the formal definition of language. Now it can be shown that when attention is restricted to language automata that are stable, receptive, reduced, and that fulfill certain completeness requirements, a glossary giving imports alone is sufficient for the reconstruction of the associated language automaton; since no two distinct language automata of this kind can have the same import glossary there is still in effect a one–one correspondence between the language automata and the glossaries. For languages of this slightly specialized class, then, imports alone are sufficient to specify meaning. By eliminating purports an economy is achieved without harmful loss of representational power.

If attention is restricted more drastically to the family of semantically structuralizable languages, a correspondingly drastic simplification in meaning representation can be made. It can be shown for a start that a purport glossary is sufficient for the reconstruction of the associated language automaton when it is known that the automaton is semantically structured. But a further simplification is possible due to the fact that a purport has under these circumstances the form of a set of sets, one member of which (the set of all modelling structures satisfying the sentence) is a superset of all the others. Clearly this one member alone is a sufficient representation of the purport, for the rest of the purport could be reconstructed from it by taking subsets. We conclude that a glossary specifying for each sentence of a language the set of models satisfying the sentence is a sufficient structuralization of the language for the meaning-theoretic study of semantically structuralizable languages. (Mathematically such a glossary is identical with the function $H[.]$.)

A minor variant of this kind of specialized glossary is obtained if the set assigned each sentence by the glossary is taken to be the set of all modelling structures *not* satisfying the sentence. Obviously, the entire language is again reconstructable from such a glossary so long as it is known to be semantically structuralizable. This variant is very close to the representation of information content proposed by Carnap (1942) and later by Carnap and Bar-Hillel (Bar-Hillel 1964, pp. 221-74). This historical circumstance illuminates the relationship between the present theory and classical theories of language: if we take the meaning-theoretic tack in logico-linguistic theory and specialize to semantically structuralizable languages, certain of the more highly compressed forms of meaning representation start to resemble some classical proposals.

It would be of interest to develop the theory of specialized glossaries further to see if it is also consistent with or could provide a more general framework for other theories of meaning that have been proposed.

6.4 Synonymy

One immediate result of the meaning-theoretic development of logico-linguistics is a natural definition of sameness of information content or synonymy. Two sentences may be said to be logico-linguistically synonymous when they have the same meaning. This is a criterion of identity of meaning rather than closeness of meaning, and is of course applicable as it stands only to whole sentences. Moreover it is synonymy in the sense of sameness of information content; two sentences could differ stylistically or connotatively and yet be synonymous under the criterion. It is not claimed to be the only interesting definition of synonymy, but it is one worthy of attention.

Does sameness of meaning make sense as a criterion of synonymy? An argument that it does can be based on the idea of behavioral indistinguishability among sentences. It turns out that sentences which are behaviorally indistinguishable in the sense that no experiment involving feeding them into a language automaton could distinguish between them are also synonymous, and conversely. This is what one would expect from sentences with the same meaning.

Sentences synonymous under the criterion of sameness of meaning are logically equivalent, but the converse does not always hold. This result is consistent with the view held by some philosophers of language to the effect that even logical equivalence, let alone such weaker relationships as material equivalence, is inadequate as a criterion of synonymy.

DEFINITION 6.5. Let s_1 and s_2 be sentences of a language $\mathfrak{L}$. Then s_1 and s_2 are *synonymous* in $\mathfrak{L}$ if and only if they have the same meaning (purport and import) in some structuralization of $\mathfrak{L}$.

DEFINITION 6.6. Let $\mathfrak{A} = \langle S, Z, L, B \rangle$ be a language automaton with s_1, $s_2 \in S$. Then s_1 is *behaviorally indistinguishable* from s_2 in $\mathfrak{A}$ if and only if for every $z \in Z$ and every pair $\bar{s}$ and $\bar{s}'$ of finite non-null sequences of sentences in S in which $\bar{s}'$ differs from $\bar{s}$ only in the substitution of instances of s_1 for s_2 and of s_2 for s_1,

$$\bar{B}(z, \bar{s}) \text{ if and only if } \bar{B}(z, \bar{s}').$$

Behavioral indistinguishability so defined is a specialization of what in automata theory is called Nerode equivalence among inputs.

THEOREM 6.1. Sentences s_1 and s_2 of a language $\mathfrak{L}$ are synonymous in $\mathfrak{L}$ if and only if they are behaviorally indistinguishable in every structuralization of $\mathfrak{L}$.

Proof. If s_1 and s_2 are synonymous in $\mathfrak{L}$ they have the same purports and imports in some structuralization of $\mathfrak{L}$ and so are behaviorally indistinguishable in it. But since behavioral indistinguishability is a behavioral property they are also behaviorally indistinguishable in every other structuralization of $\mathfrak{L}$. Conversely if s_1 and s_2 are behaviorally indistinguishable in every structuralization of $\mathfrak{L}$, then in particular they are behaviorally indistinguishable in $\mathfrak{L}$'s reduced structuralizations where their purports and imports are identical.

THEOREM 6.2.

(i) If s_1 and s_2 are synonymous sentences of a language $\mathfrak{L}$ then s_1 is logically equivalent to s_2 in $\mathfrak{L}$.
(ii) If s_1 is logically equivalent to s_2 in $\mathfrak{L}$ and $\mathfrak{L}$ is semantically structuralizable then s_1 and s_2 are synonymous in $\mathfrak{L}$.

Proof. By Theorem 5.1 and the relevant definitions.

CHAPTER 7

LANGUAGE AND INDUCTIVE LOGIC

In an earlier chapter it was seen that language and deductive logic were intimately related; that, in fact, every language determined a unique deductive logic. But what of inductive logic? We turn now to that question and the closely related problem of how to represent information believed only with a certain degree of probability.

7.1 Credibility Weights

The language automata considered up to this point have been two-valued; only the two credibility levels of belief and its absence have been distinguished. It has been assumed that a heard sentence is either received and accepted by the language automaton or not received by it at all, the input selector acting as an all-or-nothing screening device. Likewise output has been assumed to consist of yes-or-no signals to indicate whether a sentence is currently affirmable or not. These assumptions are restrictive and it is natural to generalize to the many-valued case.

Let us assume then that a set of numbers or other entities called *credibility weights* is specified as part of the language automaton. Intuitively a credibility weight signifies a degree of believability or assertability. Each informative input sentence is accompanied by a credibility weight which serves as a signal from the input selector to the language automaton instructing it as to how strongly the sentence is to be believed. Each output given in response to a test input is also a credibility weight, interpreted this time as an indication of the level of belief already accorded the sentence in the current state. The two-valued case is then the special case in which the credibility weight set contains just two weights interpreted as belief and nonbelief.

The most convenient assumption to make about the credibility weight set is that it contains a continuum of values, the interpretation being that the credibility levels range from absolute belief or certainty to absolute disbelief. Such a weight set is representable mathematically as the real line or an interval on the real line. This model, though not necessarily the most realistic psychologically, is interesting because one feels intuitively that degrees of belief are indeed graduated continuously like shades of grey.

A continuous spectrum of belief values overcomes certain behavior-analytic anomalies of the two-valued model. For example, the utterance of a sentence may be less likely when it is strongly disbelieved than when it is weakly disbelieved or when the speaker feels neutral about it, for some speakers may avoid lying more carefully than bluffing. Such output selection criteria would not be possible on the basis of mere belief–nonbelief signals from the language automaton. Similarly a preference for uttering very strongly believed sentences over those that are only fairly strongly believed could be explained only in terms of a sufficiently finely graduated credibility weight set.

In the generalized theory a language automaton becomes a *weighted language automaton* and a *weighted language* is an equivalence class of weighted language automata. The notions of a basic evidence statement, of stability, receptivity, intransigence, and so on are readily extended to weighted languages. Nonpathological languages are again stable and receptive.

A weighted language automaton has the form of a 5-tuple $\langle S, W, Z, L, C\rangle$ where S is a sentence set, W is a simply ordered set containing at least two members (the 'credibility weights'), Z is the state set, L is a (partial) function $L: Z \times S \times W \to Z$ and C (the 'credibility function') is a (partial) function $C: Z \times S \to W$. Since the members of W are indirect observables their mathematical structure may be chosen arbitrarily; taking them to be numbers is convenient in that it makes the simple ordering self-evident. Such a 5-tuple is the 'abbreviated' form of a weighted language automaton; its precise definition in the standard automata-theoretic form of Definition 2.1 is left to the reader. When translated into the standard form the standard concept of equivalence (Definition 2.5) applies to yield the definition of a weighted language.

If S is a sentence set and W a credibility weight set, a *weighted basic evidence statement* in S and W is a metalanguage statement of form

There exists (doesn't exist) a $z \in Z$ such that

$$C(\bar{L}(z, \bar{s}_1, \bar{w}_1), s_1) = w_1 \text{ and } \ldots \text{ and } C(\bar{L}(z, \bar{s}_n, \bar{w}_n), s_n = w_n$$

where $1 \leqslant n < \infty$ and for each i, $1 \leqslant i \leqslant n$, $s_i \in S$, $\bar{s}_i$ is a finite (possibly null) sequence of members of S, $w_i \in W$ and $\bar{w}_i$ is a sequence of members of W of length equal to the length of $\bar{s}_i$. The virtues of the unweighted basic evidence statements as specific behavioral properties coming as close could be hoped to providing a complete basis for language investigation carry over to the weighted basic evidence statements.

A weighted language automaton $\langle S, W, Z, L, B\rangle$ is *stable* if and only if for every $z \in Z$, $s \in S$, and $w \in W$, if $C(z, s) = w$ then $L(z, s, w) \simeq z$. It is *receptive* if and only if for every $z \in Z$, $s \in S$, and $w \in W$, $C(L(z, s, w), s) = w$. It is *intransigent* with respect to weight $w \in W$ if and only if for all $z \in Z$, $s, s' \in S$, and $w' \in W$, if $C(z, s) = w$ then $C(L(z, s', w'), s) = w$. For example, the semantically structured language automata of past chapters were intransigent with respect to the weight representing belief. Since the properties involved are behavioral all three definitions extend immediately from language automata to languages. (The generalization of the definitions to the incomplete case follows that of two-valued languages.)

A further generalization would be necessary if it were assumed that more than one credibility weight set might be needed to describe a language. For some languages it might be expedient, for example, to adopt a different weight set for the informative input than for the output, and/or different weight sets for different classes of sentences. A weighted language automaton of this fully generalized sort would have the form $\langle S, W_1^{\text{in}}, \ldots, W_m^{\text{in}}, W_1^{\text{out}}, \ldots, W_n^{\text{out}}, Z, L, C\rangle$. Whether or not more than one weight set is really needed to describe the natural languages remains to be investigated.

7.2 PROBABILITY WEIGHTS

The fundamental behavioral interpretation of the credibility weights is that they are indirect observables representing acceptability signals coming from the Input Selector and assertability signals sent to the Output Selector. No other interpretation is needed to justify postulating their existence in an analysis of verbal behavior. It may happen, however, that for the analysis of some languages or some classes of sentences within certain languages there are other interpretations consistent with the basic behavioral one. In such circumstances it is possible to say more about the credibility weights than that they are just acceptability or assertability indicators.

When the credibility weight set is taken to be the closed interval from 0 to 1, a possible additional interpretation which suggests itself is that the weights be understood to indicate *probabilities*. Thus an informative input sentence accompanied by the number 0.8 would be a signal from the Input Selector that the sentence is to be accepted as having 80% probability. Similarly the number 0.8 put out in response to a test sentence would indicate that the sentence had a current probability of 0.8 according to the language user's best information of the time. A weighted language automaton in which the credibility weights are always so interpretable will be termed a *probability-weighted language automaton* and an equivalence class of them a *probability-weighted language*. Note that a probability-weighted language automaton is just as deterministic as a two-valued one; it is analogous to a (deterministic) digital computer programmed to perform probability computations.

The probabilities in question are *personal* or *subjective* probabilities. If a language user in a certain information state believes a sentence with a credibility weight of 0.8, that figure represents *his* estimate of the probability that what the sentence says is so. The figure 0.8 is relative to his current personal information; there is nothing absolute about it. If some other language user were to estimate its probability differently, it would be meaningless to speak of one user as 'right' and the other 'wrong', for they are in different information states. It is a commonplace that different statisticians sometimes correctly compute different probabilities for the same event because each has different partial information available to him.

Experimental procedures have been developed in recent years for determining personal probabilities empirically. Techniques for eliciting personal probabilities have been used, for example, in connection with business decision-making in which the businessman's own estimates of the likelihood of various possible future events is needed in order to work out a decision that would be rational for him (Schlaiffer 1959). These techniques are available for finding out a language user's subjective probability for any sentence that can be unambiguously associated with a definite event. Such methods are the answer, or at least the beginnings of one possible answer, to the problem of finding an empirical criterion for establishing credibility levels in the course of a logico-linguistic analysis of probability-weighted languages.

To convey the flavor (though not the details, and certainly not the whole underlying utility theory) of some of the behavioral criteria that have been suggested, we quote here a typical proposal for determining personal probabilities due to Edwards, Lindman, and Savage (1963, p. 197):

> For you, now, the probability $P(A)$ of an event A is the price you would just be willing to pay in exchange for a dollar to be paid you in case A is true. Thus, rain tomorrow has probability 1/3 for you if you would pay just \$.33 now in exchange for \$1.00 payable to you in the event of rain tomorrow.

This criterion leads immediately to a very simple informant technique which, with refinements, could be used in logico-linguistic investigations. To discover for example an English informant's current credibility weight for the sentence

It will rain tomorrow, the investigator need only discover, either by questioning or direct experiment, how much the informant would be (barely) willing to pay for a certificate guaranteeing him a prize of one dollar in case the statement *It will rain tomorrow* turns out to have been correct.

For sentences such as *Brutus stabbed Caesar* or *There is life on Betelgeuse*, direct experiment is ruled out for lack of a practical criterion for deciding whether the prize promised by the certificate is to be paid or not. However, hypothetical questioning about the worth of an imaginary certificate of the required sort would still be possible. Another potential obstacle is that money does not stand in perfect linear correspondence to true subjective utility. But this obstacle can also be overcome by various refinements, one of which involves recasting the question in terms of additive 'utiles' instead of dollars, the scale of utiles being predetermined by known techniques. A related proposal for eliciting subjective probabilities which exploits the familar decision situation of whether to take a bet at stated odds is described by Stalnaker (1970).

Such behavioral criteria give for any given language user at a given time a definite personal probability for every sentence unambiguously associated with an event on which it would make sense to lay bets. There may be practical difficulties in pinpointing an exact figure, but the probability exists in principle. This has to be emphasized, because according to some philosophies of probability there exist many bettable events which simply have no probability, or for which the probability is obscure. The criteria assign probabilities to unique events, past events, and events for which the idea of a surrounding series or class of events does not make much sense. Thus at this moment I have by the criteria a definite personal probability for the sentence *The plays commonly attributed to William Shakespeare were actually written by Francis Bacon*, and this in spite of my ignorance of the historical issues involved: I would be willing to bet on it at *some* odds, and these odds determine my subjective probability for the sentence by definition.

There are of course sentences for which there is no unique bettable event – sentences for which the conditions under which the bet would be won or lost are problematic for one reason or another. In a language like English, interrogative and other non-indicative types of sentences may be candidates for this category, as are ambiguous or context-dependent sentences. Beyond this there are such sentences as *John cannot possibly lift that weight no matter how hard he tries*. Because of the difficulty of conceiving circumstances that would definitely confirm this statement, it is questionable whether it could be assigned a clear probability. A problem sentence of another sort would be *It is more probable that he will go than that she will*; since the subject matter of the sentence is itself probabilistic, one might expect difficulties in assigning a probability to it. To be conservative, therefore, the probability interpretation of continuous credibility weights should be assumed applicable only for *some* sentences, though in many languages the class of sentences to which it applies may be large. The natural languages in particular are probably not probability-weighted, though they may have substantial sublanguages that are.

Probability-weighted languages or language fragments can sometimes also be analyzed using a finite number of credibility weights interpreted as probability ranges. Thus the two-valued weight set of the earlier chapters can be given empirical meaning by establishing a probability threshold near one and defining 'belief' as a personal probability exceeding the threshold. More generally, any way of associating credibility weights with exclusive and exhaustive segments of the unit interval gives the weights an interpretation through the probability interpretation. A language automaton with a finite weight set so interpreted is mathematically a homomorphic image of the probability-weighted automaton which would have been obtained if the continuous credibility weight set had been used. If a language automaton is an accurate structuralization of a language, so are all its homomorphic images; the latter are like maps of the same territory which show less detail. The choice of whether to describe a full probability-weighted language automaton or only a homomorphic image of it is an arbitrary choice to be made by the analyst on the basis of which is easier and how finely detailed he wants his language description to be.

A practical objection which might be raised against the probability interpretation of credibility weights is that careful elicitation procedures designed to yield accurate numeric estimates of personal probabilities are apt to be delicate and time-consuming. Recall however that the informant technique consists of asking the informant whether an information state with certain properties could exist, which calls for a yes–no answer rather than a figure. In practice all the informant usually need do is to think in terms of comparisons or extreme cases. For example, if the basic evidence statement at issue is whether a rational English speaker could be so informed as to assign a probability of 0.8 to *It will be rainy and cold tomorrow* and simultaneously 0.3 to *Tomorrow it will be rainy*, the exact numbers do not matter. So long as the former number is larger than the latter no such state could exist.

The informant technique for probability-weighted languages can be described in terms of a generalized What-Do-You-Know? game. The Questioner presents informative statements to the Answerer in the form "YOU ARE NOW TO ACCEPT AT PROBABILITY LEVEL ——— THAT ———." Questions have the form "AT WHAT PROBABILITY LEVEL DO YOU BELIEVE THAT ———?" The informant must of course be trained in the intended behavioral interpretation of the probability figures. In other respects the What-Do-You-Know? informant technique is as before.

7.3 DEDUCTIVE LOGIC IN PROBABILITY-WEIGHTED LANGUAGES

A philosophical position maintained by logical empiricists and others is that *nothing is absolutely certain unless it is logically certain*. Let us call this the Philosophy of Uncertainty. You may object that there are at least some nonlogical facts of which you are absolutely certain, such as the fact that you are not in Antarctica. But are you certain with probability 1 or just practically certain with, say, probability 0.9999998? Can you think of no circumstance – self-delusion, an elaborate practical joke – which however improbable is nevertheless possible and might have caused you to be deceived in the matter? Even if you can't, does this failure of your imagination prove such circumstances don't exist? In a bet, would you be willing to give *infinite* odds against your being in Antarctica? If not you are not absolutely certain; or so the arguments for the Philosophy of Uncertainty run.

True, such locutions as "It is certain that . . ." crop up frequently in colloquial discourse. However, the colloquial usage of "certain" must not be confused with the technical sense in which it is associated with a probability of exactly 1. Cohen (1960, pp. 169-70) reports on an experiment in which 125 adults were asked to compare the use of the phrase "is certain to" with what they thought the actual intended meaning was in typical passages about politics and the weather. The average informant thought the intended probabilities were 0.73 in the political context, 0.70 in the meteorological. Evidently the English word 'certain' falls far short of expressing absolute certainty.

The Philosophy of Uncertainty suggests as a criterion of logical validity for probability-weighted languages that a sentence is logically valid if and only if it has probability exactly 1 in every information state. The 'only if' part of the criterion, according to which logical validity guarantees absolute certainty, seems plausible when it is remembered that the speakers of the language are the only final authority on the rules of their own language. If an ideal speaker of a language were to maintain that in his idiolect a certain sentence is logically valid, he could offer infinite odds on it without fear because he is the ultimate authority about his own idiolect. When you control the roulette wheel you can't lose.

In seeking a plausible criterion of logical implication it will not do to say that s_1 logically implies s_2 if and only if a probability of 1 for s_1 is always accompanied by a probability of 1 for s_2 because under the Philosophy of Uncertainty neither s_1 nor s_2 can ever have a probability of 1 if they are non-tautologous. We may require, however, that s_2 approach 1 as s_1 approaches 1, and in this way use limit operations to obtain a definition of logical implication. Logical inconsistency can be defined similarly as the inability of a set of sentences to have probabilities that approach 1 simultimateously.

To encapsulate these logical relationships and obtain some new ones it is convenient to define a generalized logical consequence relation which can have more than one conclusion sentence. This generalized logical consequence relation holds if and only if the sum of the probabilities of the conclusions must approach or exceed 1 when the probabilities of the premises all approach 1. Ordinary logical implication is then the special case in which there is only one conclusion sentence. Logical inconsistency is the case in which there are no conclusion sentences.

As an example of a new relationship, if there are no premises and the generalized implication relation holds we call the conclusion sentences 'logically disjunct'. Logical disjunctness is a relationship which is rarely mentioned in standard works on logic. Intuitively, a set of sentences is logically disjunct if it can be deduced on logical grounds alone that at least one of them must be the case. Now in languages in which classical negation cannot be expressed, logical disjunctness cannot be defined satisfactorily with just the two credibility weights belief and nonbelief. That it is readily definable in terms of probabilities is therefore to be counted an advantage of the generalization to the probability-weighted case.

A more familiar relation is that of logical contradiction, exemplified in classical logic by a sentence and its negation. Stronger than mere logical inconsistency, a contradiction may be said to exist between sentences if and only if they are not only logically inconsistent but also logically disjunct (intuitively, when there is no third possibility). Here again we have a relationship that can be defined only awkwardly in terms of belief and nonbelief, for with only these weights available it is not always possible to distinguish between logical inconsistency and logical contradiction without bringing the learning operation into the picture (cf. Examples 5.2, 5.3). But the definition is straightforward for probability-weighted languages.

To record a finding about a basic evidence statement that affirms a generalized consequence relationship, the stricture that no more than one sentence can appear below the line is lifted. One has for instance

Tom is no older than Dick
Dick is no older than Tom
Either Dick is older than Harry or Harry is older than Dick

Tom is older than Harry
Harry is older than Tom

As in the two-weighted case, most data-gathering is apt to consist of tests for basic evidence properties having the character of logical or quasi-logical relationships such at this. Because of the complicating factor of the limit operations tests for logical and quasi-logical relationships in the probability-weighted case really amount to simultaneous tests of infinite bundles of basic evidence statements, but this poses no special difficulties.

To avoid technical complications the generalized logical consequence relation will be given a precise definition here only for finite sets of sentences. This is of course no great restriction so far as the use of the concept for evidence-gathering purposes is concerned. With probability-weighted language automata represented formally as weighted language automata $\langle S, W, Z, L, C\rangle$ in which $W = [0, 1]$, the definition is formulated as follows.

DEFINITION 7.1. A finite set S_2 is a *generalized logical consequence* of a finite set S_1 in a probability-weighted language $\mathfrak{L}$ (symbolically $S_1 \models_{\mathfrak{L}} S_2$) if and only if for some (every) structuralization $\langle S, W, Z, L, C\rangle \in \mathfrak{L}$, S_1, $S_2 \subseteq S$ and for every $\epsilon > 0$ there exists $\delta > 0$ such that for every $z \in Z$ if $C(z, s) > 1 - \delta$ for all $s \in S_1$ then $\sum_{s \in S_2} C(z, s) > 1 - \epsilon$.

As special cases, if S_2 is a generalized logical consequence of S_1 in $\mathfrak{L}$ and S_2 contains exactly one member s, S_1 *logically implies* s in $\mathfrak{L}$ (abbreviated $S_1 \models s$). If moreover S_1 is empty, s is *logically valid* in $\mathfrak{L}$. If S_1 is non-empty and S_2 empty, S_1 is *logically inconsistent* and if it contains only one sentence the sentence is *logically self-contradictory*. When S_1 is empty S_2 is said to be *logically disjunct* if it contains more than one member. A set of two sentences is *logically contradictory* in $\mathfrak{L}$ if and only if it is both logically disjunct and logically inconsistent in $\mathfrak{L}$.

It is an immediate consequence of the definition that a logically valid sentence is believed with probability 1 in every state. It does not follow from the definition, however, that a logically self-contradictory sentence has probability 0 in every state.

The logical closure operation for $\models$ is a true closure operation in the algebraic sense (cf. Section 4.3). Theorem 4.3 (Section 4.4) carries over to the probability-weighted case, arguing for the naturalness of the mathematical characteristics bestowed on $\models$ by Definition 7.1 in the restricted case of no more than one conclusion sentence.

When S_2 contains exactly one sentence, Definition 7.1 is closely related to E. Adams criterion of 'reasonable inference' (1965, 1966). A slightly stronger definition would be obtained by requiring that for every $z \in Z$,

$$1 + \sum_{s \in S_1} (C(z, s) - 1) \leqslant \sum_{s \in S_2} C(z, s).$$

This strengthened version is mentioned by Stalnaker (1970, p. 66n) for the special case in which S_1 and S_2 each contain just one sentence. A weakened version would be obtained by requiring only that there exists $\epsilon > 0$ for which there is no $z \in Z$ such that $C(z, s) > 1 - \epsilon$ for all $s \in S_1$ and $C(z, s) < \epsilon$ for all $s \in S_2$. These alternative criteria are obviously not equivalent to Definition 7.1, but it is not easy to find convincing examples of interesting languages for which they produce results different from it. The question of whether they are more 'natural' or in some other way superior to Definition 7.1 remains to be investigated.

The *deductive logic* of a probability-weighted language $\mathfrak{L}$ is the pair $\langle S, \models \rangle$ where S is the sentence set of $\mathfrak{L}$ and $\models$ the generalized logical consequence relation of $\mathfrak{L}$. The idea of using generalized logical consequence as a cover concept for all the more special logical relations is implicit in Gentzen's notion of a valid 'sequent' (1935). Gentzen provided axiomatic rules for describing such a relationship for classical and intuitionistic languages, rules which incidentally suggest an approach to axiomatic language description which may be superior in many situations to the more traditional axiomatic techniques.

A dynamic implication relationship may be said to hold in case learning the premises in an indicated order with sufficiently high probability would force a high probability for the conclusion. Formally, $\bar{s}$ *dynamically implies* s in a probability-weighted language $\mathfrak{L}$ (in symbols: $\bar{s} \overrightarrow{\models}_{\mathfrak{L}} s$) if and only if for some (every) structuralization $\langle S, W, Z, L, C \rangle \in \mathfrak{L}$, $s \in S$ and $\bar{s}$ is a finite sequence of non-tautologous members of S and for every $z \in Z$ and every

$\epsilon > 0$ there exists a $\delta > 0$ such that $C(\bar{L}(z, \bar{s}, 1 - \delta), s) \geqslant 1 - \epsilon$. Strengthened and weakened versions of this definition can be stated, and as was the case with logical implication it is not yet clear which version if any is best. A dynamic inconsistency relationship is defined if '$\geqslant 1 - \epsilon$' is replaced by '$\leqslant \epsilon$'. Generalized dynamic consequence relationships can also be defined.

The definitions of the logical and quasi-logical relations for the probability-weighted case remain mathematically meaningful even when the credibility weights are not interpreted as probabilities, so long as they are still represented as numbers in the unit interval. Thus nothing prevents the definitions from being used when the numbers are interpreted simply as degrees of acceptability or assertability. The only delicate question is whether the label 'logical' is still appropriate for such relationships. We will use it in such circumstances as a technical term without prejudging the issue of its intuitive appropriateness.

7.4 THE SEMANTICS OF PROBABILITY-WEIGHTED LANGUAGES

Probability-weighted language automata can be given a semantic structure in much the same way as a binary-weighted language automaton can. The underlying precepts are still possible worlds and truth conditions, giving rise again to a structural apparatus involving a model set and a satisfaction relation. The essential difference is in the structure of the information states. Instead of a binary indication for each modelling structure of whether the corresponding state-of-affairs has been ruled out as a current possibility, the information state has now to indicate the subjective probability of that state-of-affairs in the mind of a language user in that state. An information state becomes a subjective probability measure over the model set.

In this generalized form of semantically structured automaton the model set is like an 'event space' (or 'sample space') in probability theory, the modelling structures in it being comparable to 'elementary events' ('sample points'). (In statistics an event space is sometimes described as the set of all possible outcomes of an experiment, but 'experiment' is so broadly construed that there is little difficulty reconciling the idea with the possible world notion.) The satisfaction rules associate with each sentence a compound event consisting of a set of elementary events. The language user's personal probability for a sentence at a given time is then the measure of the corresponding compound event under the current probability measure. The learning of a new sentence at a specified probability level is the changing of the current measure to a new one in which the event associated with the sentence has the indicated probability.

The Philosophy of Uncertainty imposes a restriction on the probability measures allowable as information states. Since no non-tautologous sentence must ever attain a probability of exactly one, no event corresponding to such a sentence must ever contain elementary events whose combined measure is one. This can be ensured by prescribing measures which assign zero probability to elementary events. Under this stricture a modelling structure can attain a value of one only if it is the sole member of the model set, in which case all sentences satisfied by the model are theorems in a logically complete language anyway.

A probability function which never assigns a probability of 0 to any non-empty event is said to be *regular*. It is known that a probability function is regular if and only if it has a property known as 'strict coherence'. Strict coherence has to do with rational betting behavior and it is generally viewed as a property which it would be reasonable to require of an ideal reasoner. Hence there is a strong argument for imposing the requirement of regularity, an argument that goes beyond the hazy intuitive justifications of the Philosophy of Uncertainty. The connections between regularity and strict coherence were discovered by Shimony, J. Kemeny, and R. S. Lehman and are summarized by Carnap in Carnap and Jeffrey 1971 (pp. 14-15).

A delicate and potentially controversial aspect of semantic structure in probability-weighted languages is the problem of how the learning operation should be specified. Consider a language with clear truth conditions containing sentences s_1 and s_2, and an information state for that language which assigns probabilities in the corresponding event space as shown in the Venn diagram of Figure 7.1(a). Both sentences have a current subjective probability of 0.5. Now suppose s_1 is received from the Input Selector tagged with an instruction to accept it at probability level 0.75. Should the new subjective probability measure increase the two sectors of s_1 by the same factor as in Figure 7.1(b) or might they increase by different factors as in Figure 7.1(c)?

We will assume that the change to (b) is the one that would take place in an ideal hearer. The general assumption involved may be stated as a 'Dynamic Principle of Indifference' (DPI). It complements the more familiar (static) Principle of Indifference (SPI) or Principle of Insufficient Reason as it is also called:

STATIC PRINCIPLE OF INDIFFERENCE. If among a set of mutually exclusive and exhaustive possibilities there is no reason to expect one more than any other, all are equally probable.

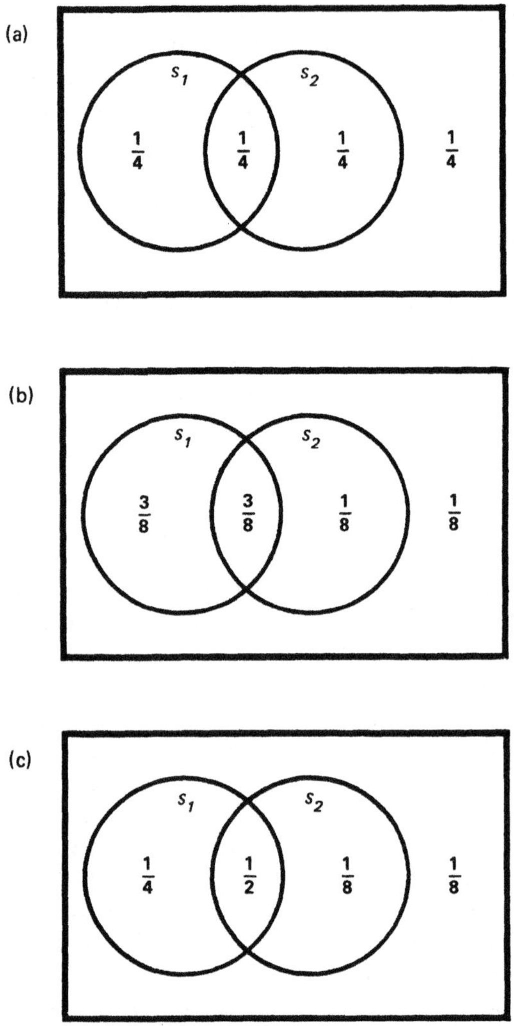

Figure 7.1. Is information state (b) or information state (c) the most rational choice of next state when the personal probability of s_1 must be raised to 3/4, starting in state (a)?

If for example a rational person knows an urn contains red, white, and blue balls, but nothing whatever about the proportions of each, he should according to SPI have a personal probability of 1/3 for the event that the next ball drawn will be red, for there is insufficient reason for him to suppose any one color to be more or less probable than any other. The dynamic counterpart of the principle is:

DYNAMIC PRINCIPLE OF INDIFFERENCE. If a set of mutually exclusive and exhaustive possibilities the probability of a subset of the possibilities is changed in the light of new information that does not affect one member of the subset more than any other, then the probabilities of the members of the subset will increase or decrease by identical ratios.

Returning to the urn example, suppose for some reason a rational person has prior probabilities of 1/3 for each of the three events: next-ball-drawn red, next-ball-drawn white, and next-ball-drawn blue. If he does not see the next draw but is told on good authority that it is not red, his personal probabilities for white and blue should each rise by the common factor of 1.5 to approximately 1/2, according to DPI.

The Static Principle of Indifference has been the object of much controversy, but the Dynamic Principle of Indifference is the only one of concern here. It says simply that ratios between probabilities of elementary events never change even when the probabilities themselves change, unless there is a definite reason to change them. If for example one possible state-of-affairs seems twice as probable as another and new information is then received that does not distinguish in any way between the two, then the one should still seem exactly twice as likely as the other though the probabilities of both may have changed.

No attempt will be made here to give a precise statement of the Dynamic Principle of Indifference. However, state changes which abide by the Principle can be described precisely as 'DPI-transforms'. Suppose S is the sentence set, M the model set, and H the satisfaction relation of a semantically structured probability-weighted language automaton. If M is finite the state set of Z of the automaton is simply the set of all functions $z: M \to (0, 1]$ such that $\sum_{m \in M} z(m) = 1$. For any states $z, z' \in Z$ and any $s \in S$, z' is a *DPI-transform of z originating in s* if and only if there exist non-negative real numbers α and β such that for every modelling structure m in $H[s]$, $z'(m) = \alpha z(m)$, and for every $m \in M \sim H[s]$, $z'(m) = \beta z(m)$. For example, the state shown in Figure 7.1(b) is a DPI-transform of the one in 7.1(a) originating in s_1 with $\alpha = 3/2$ and $\beta = 1/2$.

A compelling argument for the Dynamic Principle of Indifference is that DPI-transforms preserve statistical independence, and are moreover the only transforms that do so in general. For example, the independence of s_1 and s_2 in Figure 7.1(a) is preserved in (b) but not in (c). Intuitively, if a language user sees no connection whatever between two statements before learning that one of them is more probable than he had thought, he should see no connection between them afterward either. Several other arguments tending to establish the reasonableness of belief changes representable as DPI-transforms have been stated in some detail by Teller (1973). Teller follows Jeffrey (1965) in calling DPI-transforms a process of 'conditionalization' in which conditional probabilities remain unchanged when conditional upon the changed event; see Jeffrey (pp. 153ff., esp. 157-8) for a discussion of DPI-transforms from this point of view.

To avoid technicalities the definition of semantic structuralizability will be stated under the simplifying assumption that no model set will have to be dealt with that is not finite (or at worst, denumerably infinite). Model sets of higher cardinality can be handled but require a measure-theoretic apparatus such as that of Carnap and Jeffrey (1971).

DEFINITION 7.2.

(i) A probability-weighted language automaton $\mathfrak{A} = \langle S, W, Z, L, C\rangle$ is *semantically structured* if and only if there exists a (for simplicity, finite) set M ($\mathfrak{A}$'s 'model set') and $H \subseteq S \times M$ ($\mathfrak{A}$'s 'satisfaction relation') such that

(a) $Z = \{z \mid z : M \rightarrow (0, 1] \text{ and } \sum_{m \in M} z(m) = 1\}$;

(b) for every $z \in Z, s \in S$, and ρ such that $0 \leqslant \rho \leqslant 1$, if there exists a $z' \in Z$ such that z' is a DPI-transform of z originating in s and $\sum_{m \in H[s]} z'(m) = \rho$, then $L(z, s, \rho) = z'$ (otherwise $L(z, s, \rho)$ is undefined);

(c) for every $z \in Z$ and $s \in S$,

$$C(z, s) = \sum_{m \in H[s]} z(m).$$

(ii) A probability-weighted language $\mathfrak{L}$ is *semantically structuralizable* if and only if some structuralization of $\mathfrak{L}$ is semantically structured.

Examples of classical, intuitionistic, and modal languages that are semantically structuralizable according to this definition are easily obtained by generalizing the examples of Section 5.2 to the probability-weighted case, and a probability-weighted fragment of English is obtained by generalizing the example in Section 5.3.

THEOREM 7.1. If $\mathfrak{L}$ is a semantically structuralizable probability-weighted language, then $\mathfrak{L}$ is stable, receptive, and intransigent with respect to 0 and 1 only.

The absence of intransigence for probability weights other than 0 and 1 is one of the respects in which probability-weighted semantic theory is superior to binary-weighted. As noted earlier, rational people can change their minds about matters of fact – about sentences with subjective probabilities not equal to 0 or 1. Intransigence with respect to 0 and 1 is not objectionable because tautologies remain tautologies so long as the rules of the language remain fixed.

As the lack of general intransigence might suggest (cf. Theorem 4.6), order of premises of a dynamic implication can matter in semantically structuralizable probability-weighted languages. In classical languages, for example, $\langle \neg P, P\rangle \vec{\models} P$ but not $\langle P, \neg P\rangle \vec{\models} P$. This is what one would naturally expect, since one's most recent persuasion determines one's current beliefs. In general the order in which non-tautologous sentences are learned affects the final state of the listener unless the corresponding events are statistically independent in his mind.

The semantic definitions of the logical relationships in the probability-weighted case are similar to those of the binary-weighted case except that some further unification is achieved through the use of the generalized consequence relation.

DEFINITION 7.3. S_2 is a *generalized semantic consequence* of S_1 in a semantically structured probability-weighted language $\mathfrak{L}$ (symbolically $S_1 \models_{\mathfrak{L}} S_2$) if and only if for some (every) semantically structured structuralization $\langle S, W, Z, L, C\rangle \in \mathfrak{L}$ with satisfaction relation H, $S_1, S_2 \subseteq S$ and

$$\bigcap_{s \in S_1} H[s] \subseteq \bigcap_{s \in S_2} H[s].$$

THEOREM 7.2. ('Pragmatic Completeness'). For any semantically structuralizable probability-weighted language $\mathfrak{L}$ and any finite S_1, S_2,

$$S_1 \models_{\mathfrak{L}} S_2 \text{ if and only if } S_1 \Vdash_{\mathfrak{L}} S_2.$$

COROLLARY. Let $\mathfrak{L}$ be a semantically structuralizable probability-weighted language. Then s is logically self-contradictory in $\mathfrak{L}$ if and only if for some (every) structuralization $\langle S, W, Z, L, C\rangle \in \mathfrak{L}$, $s \in S$ and for every $z \in Z$, $C(z, s) = 0$.

7.5 PLAUSIBLE INFERENCE

In Polya's *Patterns of Plausible Inference* (1954) the following is presented as an example of an 'inductive' inference pattern:

> A implies B
> B true
> Therefore, A more credible.

It is clear from Polya's discussion that he construes such inference as taking place through time; thus the foregoing pattern asserts that when a reasoner learns B is the case, any A implying B will at the same time become more credible to him. As an English example, the pattern predicts that when someone is told *John is a radical* and accepts it, *John is a flaming radical* will become more plausible to him than it was before.

This pattern can be regarded as a metatheorem about semantically structuralizable probability-weighted languages. The ground rules are that sentences A and B are non-tautologous, that 'implies' means logically implies, that 'more credible' means 'has a higher credibility weight', i.e. is more probable, and that to learn that a sentence is 'true' means to give it a personal probability approaching 1. With these understandings Polya's inference pattern can be proved valid on the basis of definitions already introduced; specifically it can be shown that if A logically implies B in a semantically structuralizable probability-weighted language in which A is not logically self-contradictory and B not logically valid, then learning B at any credibility level higher than its current level causes the probability of A to go up also.

Table 7.1 reproduces with editorial changes Polya's chart of the fundamental inference patterns (ibid. p. 26). The pattern just discussed is the one in row I, collumn (4). *All twelve patterns are provable as metatheorems about semantically structuralizable probability-weighted languages.* Thus

TABLE 7.1

	(1) Demonstrative	(2) Shaded Demonstrative	(3) Shaded Inductive	(4) Inductive
I. Examining a Consequence	*A* implies *B* *B* false ∴ *A* false	*A* implies *B* *B* less credible ∴ *A* less credible	*A* implies *B* *B* more credible ∴ *A* somewhat more credible	*A* implies *B* *B* true ∴ *A* more credible
II. Examining a Possible Ground	*B* implies *A* *B* true ∴ *A* true	*B* implies *A* *B* more credible ∴ *A* more credible	*B* implies *A* *B* less credible ∴ *A* somewhat less credible	*B* implies *A* *B* false ∴ *A* less credible
III. Examining a Conflicting Conjecture	*A* inconsistent with *B* *B* true ∴ *A* false	*A* inconsistent with *B* *B* more credible ∴ *A* less credible	*A* inconsistent with *B* *B* less credible ∴ *A* somewhat more credible	*A* inconsistent with *B* *B* false ∴ *A* more credible

what Polya sets forth simply as natural 'heuristic' reasoning patterns are mathematical consequences of more general notions such as information states, rules of satisfaction, and DPI. Moreover, the mathematical theory is able to predict the *amount* of change in probability. In row I of column (3), for example, the chart says only that A will become "somewhat more credible" whereas with DPI and the other assumptions implicit in semantic structuralizability one is able to predict that A's probability will rise in the same proportion as B's.

Interesting applications of the inference patterns of Table 7.1 arise in the methodology of the sciences. The hypothetico-deductive method as described by Popper and others (Section 3.3) is essentially an application of the patterns in columns (1) and (4) of row I: letting A be the scientific hypothesis and B one of its observational consequences, the 'demonstrative' pattern in column (1) describes the rejection of the hypothesis because of falsification by the evidence of its observational consequence, while the 'inductive' pattern of column (4) describes an increase in confidence in the hypothesis due to empirical verification of the observational consequence. When the science in question happens to be logico-linguistics, the hypothesis to be tested is a language description and the observational consequence a basic evidence statement.

EXAMPLE 7.1. *An Invalid Inference Pattern.* In addition to the twelve in the table, Polya gives some further examples of plausible inference patterns one of which is the following (ibid. p. 27):

> H implies A.
> H implies B.
> B true.
> Therefore A more credible.

Though it seems plausible enough at first, this inference schema can *not* be derived from our definitions, and by examining the point at which the proof breaks down it is possible to construct counterexamples to it. Imagine for instance a raffle in which a random draw has been made from a bowl containing tickets numbered 1 through 1000 and the drawn ticket is about to be read. Let A be *The number drawn is even*, let B be *The number drawn is prime*, and let H be *The number drawn is* 2. Upon being told B an onlooker would find A *less* credible than before so the inference pattern fails. This example is not intended to disparage in any way Polya's pioneering work, but only to underscore the prophylactic value of having a detailed formal theory to serve as a check on one's logical intuition.

7.6 STATISTICAL INFERENCE

It would be out of the question to attempt to explore here the full connection between logico-linguistics and all branches of statistics. However, there may be an especially intimate link between logico-linguistics and Bayesian probability estimation and hypothesis testing. The Bayesian statistician is interested in how his 'prior probabilities' for a hypothesis should be affected by observing the result of a sampling process; he sees statistics as the theory of how to change these prior probabilities to 'posterior probabilities' that take the new evidence into account in a rational way. Thus the Bayesian statistician, like the logico-linguist, is a student of information state changes. More specifically, he is like a logico-linguist investigating semantically structuralizable probability-weighted languages in that he considers subjective probability distributions adequate as representations of information states for his purposes.

In statistics as commonly practiced there is no emphasis on the way in which hypotheses of interest are expressed linguistically, and this is a point of difference between statistics and logico-linguistics. But it is only a difference in emphasis, for the statistician is normally obliged to describe in some way all events of concern to him, and if they cannot be expressed adequately in English he resorts to a formal language or invents his own symbolism to name them. Once named, the events are studied by the Bayesian statistician in much the same way as they would be by a logico-linguist specializing in the effects of the learning operation in the semantically structuralizable probability-weighted case. The statistician is in fact an import specialist (cf. Section 6.1), for the way in which learning a new observation sentence changes prior probability distributions into posterior probability distributions is, logico-linguistically, just the import of the sentence.

From the Bayesian viewpoint, simple classical hypothesis testing is a partial account of Bayesian hypothesis testing in which, in order to avoid assumptions about the prior probabilities, the results of the test are reported as the confidence level at which the 'null' hypothesis may be rejected. If this view is accepted much of classical statistics fit into the logico-linguistic scheme of things as a part of Bayesian statistics. But lest it be thought that every logico-linguist must be an expert statistician, it should be pointed out that most statistical complexity occurs at the stage at which the statistician attempts to specify a reasonable prior probability distribution. The inference mechanism of statistics is the same as the logico-linguistic one, but the logico-linguist interested in language mainly as a vehicle for communication need not ordinarily worry about how to calculate the exact probabilities in the prior distribution.

EXAMPLE 7.2. *Comparison of Two Simple Hypotheses.* An elementary statistics textbook (Bryant 1966, p. 37 ff.) describes the problem of a Mr. Bone who raises chickens for the market. Bone is contemplating the purchase of 1,000 baby chicks. He learns from the hatchery operator that the 1,000 chicks represent either a mixture of 250 males and 750 females or else a mixture of 750 males and 250 females. The hatches have gotten mixed up; the operator knows it is one mixture or the other but has no idea which. Mr. Bone is granted the privilege of selecting 5 chicks at random from the lot of 1,000 to determine their sex. He does so and all five chicks in his sample turn out to be male. What can he deduce about which mixture of males and females the lot contains?

As a classical statistician would view the problem, there are two simple ('exact') alternative hypotheses to be considered. They are

H_0: The lot contains 250 males and 750 females;

H_1: The lot contains 750 males and 250 females.

Also, an observation has been made which may be stated

O: A random sample of 5 chicks turned out to be all male.

By a routine computation (using the binomial approximation to the hypergeometric distribution) one obtains

$$P(O/H_0) = (\tfrac{1}{4})^5 = 1/1{,}024$$

as the probability of getting the observed result O in case H_0 is true. At this point a classical statistician would probably declare hypothesis H_0 to be rejected, for under H_0 the observed result has probability small enough to warrant rejecting H_0 at the 0.001 significance level. In classical statistics the solution of the problem would be left in this form.

A Bayesian statistician, on the other hand, would not attempt to draw any final conclusion from the one conditional probability $P(O/H_0)$, which to him is of no great interest in and of itself. To the Bayesian, the important thing is to examine what Bone's personal probabilities for H_0 and H_1 should be before and especially after learning O – his 'prior' and 'posterior' probabilities for H_0 and H_1. Suppose for lack of any indication that one mixture of chicks is any likelier than the other, Mr. Bone's prior probabilities are assumed (by SPI) to be

$$P(H_0) = P(H_1) = \tfrac{1}{2}.$$

To obtain his posterior probabilities one first computes the other conditional probability

$$P(O/H_1) = (3/4)^5 = 243/1{,}024$$

and then applies Bayes' Theorem to get

$$P(H_0/O) = 1/244$$

and similarly

$$P(H_1/O) = 243/244.$$

In other words Mr. Bone ends up, if he reasons correctly, favoring H_1 over H_0 by odds of 243 to 1. To the Bayesian this seems a more meaningful way of describing the situation than the classical statistician's conclusion that H_0 is rejected at the 0.001 level. After all, it is the relative credibility of the two hypotheses that Mr. Bone is really interested in.

Now let us re-examine the situation from the logico-linguistic standpoint. Suppose Mr. Bone already speaks, or for the sake of his problem is able to construct, a semantically structuralizable probability-weighted language capable of expressing the events H_0, H_1, and O. Knowing that the sample of 5 has been drawn but prior to learning the result of the tests, his information state is that of Figure 7.2(a). There H_0 and H_1 are represented as logically incompatible sentences each having probability just under 0.5 (not exactly 0.5 because it is logically possible the operator is mistaken and neither mixture is the actual one). If Bone's prior information state is reasonable in the sense of being consistent with standard probability computations, statement O will have prior probability just under 1/2,048 jointly with H_0 (since $P(O, H_0) = P(O/H_0)P(H_0) = (1/1{,}024) \cdot \tfrac{1}{2}$) and just under 243/2048 jointly with H_1, as shown. When Bone learns the result of the test – that is, when he accepts statement O at a credibility level near 1 – he changes his state to the posterior distribution of Figure 7.2(b). By virtue of DPI the ratios within O remain unchanged, so Bone ends up favoring H_1 at a probability level of approximately 243/244. Thus the logico-linguist predicting the state change in an ideal user of a semantically structured probability-weighted language arrives eventually at approximately the same

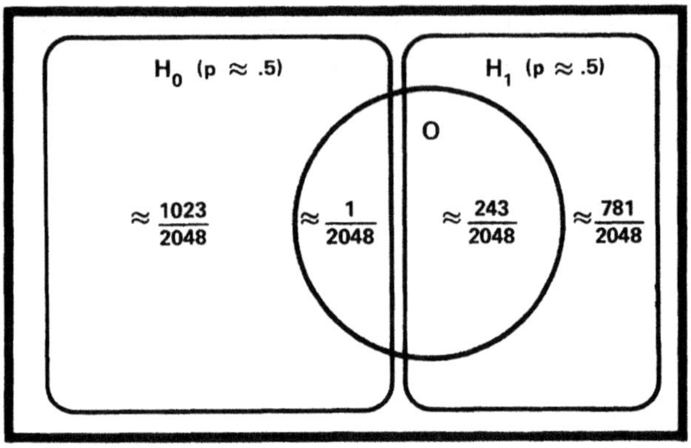

(a)

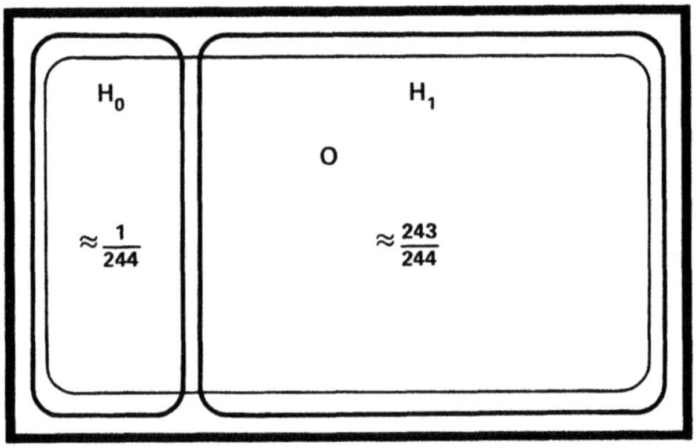

(b)

Figure 7.2. Mr. Bone's personal probability distributions (a) prior to and (b) after learning O.

result as the Bayesian statistician, provided he goes through as detailed a computation of the prior distribution.

The relationship of Bayesian statistics to logico-linguistics is obscured somewhat by notational differences. In the Bone problem, the results of the conventional Bayesian analysis would normally be presented in the form of the conditional probabilities $P(H_0/O) = 1/244$ and $P(H_1/O) = 243/244$. These are sometimes discussed as though they were the posterior probabilities, but actually they are as much prior as posterior since they are as true of Figure 7.2(a) as 7.2(b). A clearer notation would result if conditional probabilities were abandoned entirely when stating final results, indicating instead with subscripts or some other device the state referred to. Then all that would have to be said would be that in the prior state z_1 the probabilities of interest are $P_1(H_0) = P_1(H_1) \approx 0.5$, while in the posterior state z_2 they are $P_2(H_0) \approx 1/244$ and $P_2(H_1) \approx 243/244$. See Suppes (1966) for a notation akin to this and an argument that its use immediately clears up certain well-known paradoxes of statistical inference.

7.7 Inductive Reasoning

Does the study of languages, and in particular semantically structuralizable probability-weighted languages, subsume the study of inductive logic? It depends, of course, on what is meant by 'inductive logic'. If that phrase is taken to mean the kinds of plausible and statistical reasoning just discussed, the answer is obviously in the affirmative. More generally, logico-linguistic pragmatics includes all of inductive logic when inductive logic is understood to be the study of changes brought about in personal probability distributions by the learning of linguistically expressible evidence. This is not far from the sense in which the phrase is used by e.g. Carnap and Jeffrey (1971).

For convenience let us lump both deductive and inductive logic together under the single cover term 'logic'. Might there be an abstract entity which could serve as a mathematical representation of a logic, both deductive and inductive? On attempting to discover such a representation, one finds that so many parts of the language automaton are so intimately involved with either deductive or inductive logic that it seems doubtful whether any substantially simpler representation of a 'logic' could be found than the language automaton itself or its equivalence class. Deductive logics can be represented more simply, but the representation of an entire deductive–inductive logic may as well be taken to be the same as the representation of a language. This line of thought leads to a strengthened version of the Logico-Linguistic Thesis:

LOGICO-LINGUISTIC THESIS. Linguistics (beyond syntax but exclusive of higher pragmatics) and logic are concerned with essentially the same subject matter.

7.8 EXTENDED SEMANTICS

The language property of being semantically structuralizable and probability-weighted was defined in such a way as to require: (i) that the language have a clear-cut satisfaction relation, with every sentence having definite truth conditions; and (ii) that all credibility weights be interpretable as probabilities associated with rational betting behavior. Let us call a language of this sort semantically structuralizable and probability-weighted 'in the narrow sense'. The question arises: Are all interesting languages semantically structuralizable and probability-weighted in the narrow sense?

One can readily construct imaginary languages which are not. Imagine for example a linguistic community whose major social activity is wagering. It is not inconceivable that the language used by such a community might evolve in such a way as to include many devices for talking about personal probabilities; for example, it might contain a sentence connective '$*$' such that a person uttering '$s_1 * s_2$' is understood to be asserting that the probability of s_1 is for him very different from the probability of s_2. The precise rule might be that '$s_1 * s_2$' becomes assertible when the absolute difference between the speaker's personal probabilities for s_1 and for s_2 becomes sufficiently high – the closer to 1 the more assertible. By this rule the credibility weight of '$s_1 * s_2$' is a number between 0 and 1 but it is not really a probability; determing who had won a bet on it would be problematic, for example. One can imagine a language equipped with '$*$' being used successfully to describe and alter information states and hence to convey information, but it would not be a sematically structuralizable probability-weighted language in the narrow sense.

Now a difference between probabilities is not itself a probability, but it is a function of probabilities. Languages containing devices such as $*$ may therefore be accommodated by relaxing the requirements on semantically structuralizable probability-weighted languages to read: (i) there is a clear-cut satisfaction relation for at least *some* sentences (let's call them the 'bettable' sentences); (ii) the credibility weights of bettable sentences are interpretable as probabilities; and (iii) all other sentences have credibility weights which are functions of bettable sentences nested within them. Let us call a language fulfilling these weakened requirements semantically structuralizable and probability-weighted *in the extended sense*, and the semantic analysis of such a language *extended semantics*.

No precise definition will be attempted for the family of languages that are semantically structuralizable and probability-weighted in the extended sense, but some informal remarks can be made abouts its general character.

It is obvious that a language of this kind has a sublanguage which is semantically structuralizable and probability-weighted in the narrow sense, namely the sublanguage of all bettable sentences. In any structuralization the state set of this sublanguage is also the state set of the whole language, so nothing new is called for in the way of information state representations. Sentences outside the bettable sublanguage may still have as credibility weights real numbers ranging from 0 to 1, but such weights do not indicate ordinary probabilities but functions of probabilities best interpreted as mere degrees of assertability or acceptability.

In the next chapter some evidence is examined which suggests that *English is at best semantically structuralizable in the extended sense*. The main evidence concerns the English *if–then* connective in its indicative usages. A sentence of form *If* s_1 *then* s_2 has a credibility weight which is apparently best explained as the ratio of the joint probability of s_1 and s_2 to the probability of s_1 provided s_1 and s_2 are bettable (this weight is interpretable as a *conditional* probability), and as a more complex function of probabilities when s_1 and s_2 are themselves compound statements containing *if–then*. If this evidence is any indication, the appropriate kind of semantics for the analysis of natural languages is extended semantics.

CHAPTER 8

'IF-THEN': A CASE STUDY IN LOGICO-LINGUISTIC ANALYSIS

> "In this very thing, which the dialecticians teach among the elements of their art, how one ought to judge whether an argument be true or false which is connected in this manner, 'If it is day, it shines', how great a contest there is; – Diodorus has one opinion, Philo another, Chrysippus a third. Need I say more?"
>
> Cicero

Can logico-lingustic theory actually be applied in a practical way to analyze and explain significant natural-language phenomena? As an illustration of how the theory reduces to practice a case study will now be made consisting of a detailed logico-linguistic analysis of a particular natural-language device. The device selected for the study is the *if–then* construction of everyday English.

It will be seen that *if–then* in its indicative usages provides an admirable testing ground for logico-linguistic theory in that the theory's features of probability weighting, extended semantics, and the distinction between logical and dynamic implication and inconsistency are all needed to formulate and test interesting hypotheses about *if–then*. Moreover, *if–then* has historically been of great interest to those concerned with the nature of human reasoning in general and the reasoning of English speakers in particular. Consequently there is an extensive philosophical literature about it to be drawn upon, some of which suggests ideas for hypotheses and even contains wisps of logico-linguistic evidence.

8.1 PRELIMINARY STATEMENT OF HYPOTHESES TO BE TESTED

To illustrate the process of formulating and testing a logico-linguistic hypothesis it would perhaps be sufficient to treat thoroughly just one interesting hypothesis. However, there is value in being able to compare one hypothesis against another, and for that reason two competing hypothesis about *if–then* will be examined here. Both are about the naive colloquial usage of *if–then* by the man on the street as opposed to the logician's or mathematician's technical usages of conditionals in proofs. Both hypotheses are concerned with *if–then* constructions in the indicative mood and their interplay with *not*, *and*, and *or* used as sentence operators and connectives.

The first hypothesis can be stated crudely as follows:

Hypothesis A. English *if–then* has the properties of the truth-functional ('material') conditional.

This hypothesis says essentially that *if–then* is formalizable or translatable as the material conditional of classical logic, as most elementary logic books seem to suggest.

The second hypothesis associates *If* s_1 *then* s_2 with the probability of s_2 given s_1, i.e. with a conditional probability. At least, this is what the hypothesis claims when s_1 and s_2 do not themselves contain any instances of *if–then*; when they do the conditional probability idea has to be extended and the rules of assertability are less simple (the precise rules will be stated later). The second hypothesis can therefore be stated in preliminary suggestive form as follows:

Hypothesis B. English *if–then* has the properties of an extended conditional probability connective.

Since Hypothesis B differs from Hypothesis A, both cannot be correct, though both could be wrong. The aim is to test them both out to see which, if either, accords with the facts of English usage.

8.2 History of Hypothesis A

The notion that the conditional of ordinary discourse is truth-functional has a long and venerable history. In antiquity Philo of Megara stated the claim very clearly concerning the conditional connective εἰ of ancient Greek. Here is Philo's doctrine as recorded by Sextus Empiricus (trans. 1935, p. 297):

Thus Philo declared that 'the hypothetical is true whenever it does not begin with what is true and end with what is false'; so that, according to him, the hypothetical is true in three ways and false in one way. For whenever it begins with truth and ends in truth it is true, as thus – 'If it is day, it is light'. And whenever it begins with what is false and ends in what is false, once more it is true, as for instance 'If the earth flies, the earth has wings'. Likewise also that which begins with what is false and ends with what is true is true, as thus – 'If the earth flies, the earth exists'. And it is false only in this one way, when it begins with truth and ends in what is false, as in a proposition of this kind – 'If it is day, it is night'; for when it is day the clause 'It is day' is true, and this was the antecedent, but the clause 'It is night', which was the consequent, is false.

It is well known that the truth-functional approach to logic exemplified in this passage leads to the classical logic of Frege, Russell, and Whitehead whose conditional is the material conditional. Hence if this report of his opinions is to be trusted, Philo may be regarded as the spiritual father of Hypothesis A.

Modern logic textbooks follow Philo's lead with varying degrees of assuredness. A few are commendably cautious; for example in Church's *Introduction to Mathematical Logic* the English *if–then* is said only to be a 'convenient oral reading' for the material conditional, and this remark is accompanied by the disclaimer, "... these oral readings must not be taken as indicating that the meanings of these English words are faithfully rendered by the corresponding connectives in all, or even in most, cases ..." (1956, p. 37). Other texts seems to take Hypothesis A more for granted; thus Quine writes in an early textbook that "... the indicative conditional can always be construed truth-functionally, *viz.* as false just in case its 'if'-component is true and its 'then'-component false, even though its affirmation will ordinarily be motivated by considerations of causal connection ..." (1940 p. 25). Many textbooks also assign exercises in translating between formal-language sentences and English, driving home in this way the alleged interchangeability of the material conditional and *if–then*. Kalish and Montague's *Logic: Techniques of Formal Reasoning* even goes so far as to list explicit algorithmic *rules* by which this translation can be carried out in either direction (1964, pp. 4-13).

Ingenious arguments are sometimes given for why *if–then* should be considered truth-functional. An example is the one in Faris' *Truth-Functional Logic* (1962), which attempts to show the interderivability of *If* s_1 *then* s_2 with '$s_1 \supset s_2$'. The earliest argument of this kind of which I am aware is a lengthy one given by C.I. Lewis (1918 pp. 223-6) in the form of a theorem purporting to prove mathematically (!) the equivalence of English *if–then* with the material conditional. There is insufficient space here to comment on these arguments except to point out that since they make no essential use of evidence about actual usage it is hard to take them seriously.

The philosophical journal literature on the subject is mixed, with some authors apparently favoring Hypothesis A and others opposed. An example of an author in favor is L. Simons, who presents (1965) a detailed argument intended to prove from intuitively acceptable assumptions that the material conditional is indeed "the important conditional which occurs in both ordinary and scientific discourse". Other authors oppose the hypothesis, often presenting interesting alternative hypotheses in its place. The alternatives are unfortunately far too numerous to be discussed individually here, but it is notable that in most cases only one or two examples of actual English usage are offered in support of them.

Many mathematical logicians, though not ordinarily concerned with the relationship of their formal systems to colloquial usage, are willing to give offhand opinions on the subject when asked. These opinions too seem to be mixed. I have had the experience of being told in effect, "Aside from a few differences of stylistic detail, the English conditional has the same logical characteristics as the material conditional, and I don't think anyone who knew his logic would tell you otherwise"; and of being assured later by another source, "Of course the material conditional is nothing like *if–then*, it was never intended to be, and I don't think any qualified logician would tell you otherwise." Interesting intermediate positions are sometimes encountered too. I have heard a prominent metamathematician express his opinion (though not for attribution) that although the present colloquial English usage of *if–then* still differs somewhat from the material conditional, this situation might well change in another fifty years or so when the general public becomes better educated in the uses of symbolic logic. He added that he felt his own personal usage of *if–then* had already come to reflect the properties of the material conditional.

This is of course only anecdotal evidence about casual opinions, but it suggests that the disparity of opinion about the material conditional which started in ancient Greece and pervades the literature is still with us.

8.3 History of Hypothesis B

So far as I know there are no ancient teachings which imply Hypothesis B with anything like the clarity with which Philo stated Hypothesis A. Perhaps the ancients did not attach enough importance to the notion of information states, especially states of partial informedness, to be led to Hypothesis B. Charles Sanders Peirce wrote (1885, p. 280):

. . . that is possible which, in a certain state of information, is not known to be false. By varying the supposed state of information all the varieties of possibility are obtained. Thus *essential* possibility is that which supposes nothing to be known except logical rules. *Substantive* possibility, on the other hand, supposes a state of omniscience. Now the Philonian logicians have always insisted on beginning the study of conditional propositions by considering what such a proposition means in a state of omniscience; and the Diodorans have, perhaps not very adroitly, commonly assented to this order of procedure . . .

Hypothesis B does what Peirce suggests, and what both Philoneans and Diodorans failed to do, by considering states of information in which knowledge is only partial (i.e. probabilistic) and in which in particular the antecedent of a conditional may be neither known with certainty to be true nor known with certainty to be false. The foregoing passage with its explicit mention of 'information states' makes Peirce a candidate for spiritual father of Hypothesis B.

After probability theory had been formalized it must have occured to many people to associate *if–then* statements with conditional probabilities, but the connection does not seem to have been investigated seriously prior to the paper on the subject published by E. Adams in 1965. In it and in later papers (1965, 1966, and 1975) Adams constructed examples of reasoning in ordinary English supporting a relationship between *if–then* and conditional probabilities and investigated some of the formal properties of conditional probability connectives. At about the same time as Adams' early work, R.C. Jeffrey suggested such a relationship independently (1964). More recent work relating *if–then* to conditional probabilities in one way or another includes that of R.C. Stalnaker (e.g. 1970) and B.D. Ellis (1973 and forthcoming). David Lewis has demonstrated the difficulty of reconciling the idea of a conditional probability connective with conventional rules of satisfaction (1976).

Hypothesis B, in its detailed statement to be given presently, adapts Adam's and Jeffrey's concept of a conditional probability connective to the theoretical framework of logico-linguistics and extends them along lines anticipated in Cooper (1968). Specifically, Hypothesis B modifies the idea of a conditional probability connective in such a way as to take into account the learning operation as well as the credibility function, and extends it to handle sentences in which *if–then* occurs other than as the main connective. Of the English arguments to be examined as evidence, several are due to Adams (esp. his 1964), several others are taken from Cooper (1968), a few are examples of C.L. Stevenson (1970), and the rest are new. A hypothesis somewhat similar to our Hypothesis B was recently discovered independently by B. Skyrms.

8.4 HISTORY OF OTHER HYPOTHESES

Many other hypotheses about *if–then* have been put forward at one time or another. Claims are sometimes heard for strict implication, for the intuitionist conditional, and for conditionals arising from certain well-known systems of three-valued and many-valued logic. Many interesting nonstandard conditionals have also been associated with *if–then*; among them are those proposed by N. Rescher (1962) and by A.R. Anderson and N.D. Belnap, Jr. (1962). Some systems of logic involving novel conditional connectives are of course intended only as instruments of abstract reasoning, while others are presented in such a way as to suggest that they capture the properties of the English *if–then*. It is in this latter role of hypotheses about *if–then* that such systems are of interest in the logico-linguistic analysis of English. None will be discussed explicitly in what follows, but readers who happen to have a special interest in one or another of them may wish to bear them in mind and test them out against the linguistic evidence as it is reviewed.

To sum up, the current status of the *if–then* puzzle might best be described as one involving many hypotheses, much controversy, and few empirical facts. For the serious language scientist whose aim is to resolve some of this controversy rather than to add further fuel to it, the only acceptable course is to make sure the hypotheses of interest have been articulated clearly and then to say calmly, "Let us look at the evidence".

8.5 DELINEATION OF CONSTRUCTIONS OF INTEREST

By the *if–then* connective we mean the English sentence compounding device involved in statements of the form '*If* . . . *then* _____', where . . . and _____ are themselves well-formed English statements. The constructions '*If* . . . , _____' and '_____ *if* . . .' will be assumed to be stylistic variants of the same connective. Related constructions such as '*Since* . . . , _____', '*Assuming* . . . , _____', '*Suppose* . . . ; *then* _____', and '_____, *provided that* . . .' are sometimes thought to be further stylistic variants of the *if–then* construction, but their equivalence with *if–then* is questionable; in any case they will not be taken up here. The special usages of the word '*if*' in the contexts '*only if*', '*even if*', and '*as if*' will also be excluded from consideration, along with the '*whether*' usages of '*if*' (e.g. *I wonder if it will rain*). Attention will be focused exclusively on the straightforward usages of *if–then* usually discussed in logic textbooks (e.g. *If it rains, then we won't go*).

Attention is further restricted to *if–then* constructions that are in the indicative mood (e.g. *If he was there, he saw it*) as opposed to the various forms of the subjunctive (*If he had been there, he would have seen it, He would have seen it had he been there*, etc.) Although claims have been made to the effect that there are no important differences, it has in my opinion been clearly established by Adams (1970), as well as by Chisholm, Goodman, Rescher, and other earlier authors, that subjunctive conditionals possess peculiar logical properties which distinguish them from indicative conditionals. For present purposes, the subjunctive conditional can be regarded simply as a separate connective which is arbitrarily excluded from our analysis.

It is sometimes asserted that the subjunctive conditional is more important than the indicative, but if 'more important' means more frequently used, this does not seem to be the case. In a corpus of 325 occurrences of *if–then* which the author collected in 1962 by searching the text of scientific and other publications by computer, 295 (i.e. 91%) turned out to be in the indicative mood.

It is commonly assumed that the subjunctive uses of *if–then* are identical with its 'counterfactual' usages and *vice versa*, but this assumption is probably unjustified. Apparent counterexamples to it have been given in the literature in the form of subjunctive usages of *if–then* in which the antecedent is not known to be false (Burks 1951, pp. 366-7; Chisholm 1946, p. 483). But however that may be, the criterion of inclusion or exclusion for the present study is the purely syntactic criterion of grammatical mood.

In addition to *if–then*, we will also be concerned with the three other English particles *and*, *or*, and *not*. The particle *and* will be of concern only when used as a sentence connective (e.g. in *Jack fought and John fought* but not in *Jack and John fought*). It will be assumed to have as an equivalent stylistic variant the construction '*Both ... and* ______'. Uses of *and* which imply a temporal order (e.g. *John undressed and he got into bed*) will be excluded from consideration. The *or* particle will also be of concern only as a connective of whole sentences. As working hypotheses, *either–or* and *or else* will be assumed tentatively to be stylistic variants of *or*. They are sometimes said to have a less inclusive sense than *or* alone, but since to my knowledge that claim has never been systematically investigated the question will simply be left open here. The particle *not* is again of interest only as a modifier of whole sentences (e.g. in *John is not able to do it* but not in *John is not only able but willing to do it*). An informal way of seeing whether *not* is being used to modify the whole sentence is to see whether meaning is preserved when the *not* is removed and the sentence is preceded by the locution '*It is not the case that* _____'.

As will be discussed in more detail in the next chapter, an informant can make judgements about basic evidence properties only after he has interpreted a sentence, or been given an interpretation of it. This means in particular that syntactic ambiguities must be broken for him. For example, the sentence

> *Either John will be there or Harry will be there and John's wife will stay away*

is ambiguous and cannot be understood in isolation unless a clue is given to how it is to be interpreted, as for instance by the bracketing

> *Either John will be there or [Harry will be there and John's wife will stay away]*.

In considering sentences that are ambiguous in this way, we will simply insert parentheses to indicate the interpretation so that the reader who is acting as his own informant will be able to disambiguate them in the intended way. We will also assume that the intended meanings of pronouns and other deictic expressions are clear enough for practical purposes in the examples to be considered in this chapter, so that no special notation is needed to clarify them.

A special aspect of the interpretive problem which complicates the analysis of *if–then* sentences is the possibility of implicit variables linking antecedent and consequent. This is well illustrated by an example due to Burks (1951). Consider the apparently straightforward sentence

> *If it rains he'll wear his raincoat.*

The contrapositive of this statement, which by classical logic is supposed to say the same thing, is the absurd conclusion

> *If he doesn't wear his raincoat it won't rain.*

Part of the problem, at least, is that there is a hidden time variable linking the antecedent to the consequent. When the hidden variable is made explicit, the problem largely disappears. Thus

> *If it rains* (when he looks out the window) *he'll wear his raincoat* (at the later time when he goes out the door)

has as its contrapositive

If he doesn't wear his raincoat (when he goes out the door) *it won't have been raining* (at the earlier time when he looked out the window)

which doesn't seem so unreasonable. The moral is that hidden variables linking antecedent and consequent must be guarded against, either by making the variables explicit or by excluding such problematic constructions from consideration entirely.

8.6 THE WORKING HYPOTHESIS OF EXTENDED SEMANTIC STRUCTURALIZABILITY

One kind of linguistic study consists in the analysis of a small fragment of a natural language in full detail, leaving nothing unsaid about the internal structure of any of the sentences in the fragment. The task at hand is not like this, however. The goal is to analyze instead certain ways in which sentences can be joined together to make longer sentences, where the inner structure of the original sentences is to remain unanalyzed. It is necessary, therefore, to make some kind of tentative working assumption about the primative sentences from which the longer sentences are to be built up. The working assumption which will be taken as the starting point of both hypotheses is that the primitive sentences belong to a semantically structuralizable probability-weighted sublanguage of English.

This assumption may or may not be correct, but some such working hypothesis is needed as a starting point for the analysis of the connectives. Stated more fully the assumption is this. It is supposed that there exist a large number of sentences of English which taken together make up a semantically structuralizable probabilistically weighted sublanguage of English. This sublanguage will not be delineated precisely, but it will be presumed to include most of the simpler English statements one is likely to come across. Sentences containing *if–then*, *and*, *or*, or *not* are specifically excluded from this sublanguage, since we want it to include only the simple statements out of which the compound sentences that use these connectives are to be formed. Sentences containing other problematic connectives such as *since*, *unless*, etc., are also excluded, as are sentences which have a modal character (e.g. *John is probably there*, *It may not be certain yet*), since it is not clear that these are semantically structuralizable. To simplify things further we will assume that non-declarative sentence types and sentences displaying a high degree of vagueness are excluded as well. We will call the sentences in this sublanguage *simple sentences* and use the letters a, b, c, and d as metavariables ranging over them.

The proper subject matter of our investigation is then the set S of all sentences which can be built up out of the simple sentences in this sublanguage by means of the connectives *if–then*, *and*, *or*, and *not*. The sentences in this larger set will be indicated with the metavariables s, s_1, s_2, s_3, etc. Those that are not simple will be called *compound*. Thus any sentence a from the semantically structuralizable sublanguage is a (simple) sentence s in this larger set, and if s_1 and s_2 are in the set so are the compound sentences *If* s_1 *then* s_2, s_1 *and* s_2, s_1 *or* s_2, and *not*-s_1. Examples of such sentences would include English sentences of the form *a and b*, *If* [*a and b*] *then c*, and *If a then* [*if b then* [*c and not-a*]]. Notice that the sentence set S which is to be our linguistic universe does not contain all the sentences of English, but only the simple ones in the trouble-free semantically structuralizable sublanguage plus others formable out of these using the four connectives of interest. Notice too that S as a whole is not assumed to be semantically structuralizable; whether or not S is semantically structuralizable is one of the things to be investigated.

8.7 Exact Statement of Hypothesis A

Having stated some of the underlying assumptions common to both hypotheses to be considered, we may now formulate Hypothesis A more precisely. The sublanguage of English made up of the simple sentences, being by assumption semantically structuralizable, can be described structurally in terms of a model set and a satisfaction relation relating these simple sentences to the modelling structures in the model set. Hypothesis A says simply that this satisfaction relation can be extended to the compound sentences in the standard manner familiar to classical logicians. Thus Hypothesis A comprises the following set of assumptions about the satisfaction relation:

A1. A sentence of form *if* s_1 *then* s_2 is satisfied by a modelling structure m if and only if either s_1 is not satisfied by m or s_2 is satisfied by m (or both).

A2. A sentence of form s_1 *and* s_2 is satisfied by m if and only if s_1 is satisfied by m and s_2 is satisfied by m.

A3. A sentence of form s_1 *or* s_2 is satisfied by m if and only if either s_1 is satisfied by m or s_2 is satisfied by m or both.

A4. A sentence of form *not-s* is satisfied by *m* if and only if *s* is not satisfied by *m*.

These rules are easily formalized after the fashion of Example 5.4 (p. 102 line (3a)) to yield precise structural rules of satisfaction. They extend the satisfaction relation to all members of the sentence set *S* and in so doing specify a semantically structured structuralization for a language whose sentence set is *S*.

Hypothesis A claims that the language so described is part of English. It says that a structural language description for *S* of which A1-A4 are the lines dealing with the connectives would, if it had in addition to A1-A4 accurate rules of satisfaction for the simple sentences, be an accurate description of the sublanguage of English whose sentence set is *S*. It postulates that this language is semantically structuralizable, which is to say that the learning operation and credibility function are assumed to be determined by the satisfaction relation as in any other semantically structuralizable probability-weighted language. One of the obvious implications of Hypothesis A is that the sentential logic of the sublanguage of English with sentence set *S* is just the classical logic of material implication, conjunction, disjunction, and negation. The question to be explored empirically is: Is the sentential logic of this portion of English really classical in this way?

The metalinguistic (unitalicized) sentence connectives used in A1-A4 are understood to be in 'mathematician's English'. Thus the logic of the metalanguage in which Hypothesis A and B are expressed is classical, and 'if–then' in the metalanguage may be interpreted as the material conditional. There is nothing strange or suspicious about this. It is natural to choose a metalanguage whose logic is already well-known so that there can be no question about what observational consequences can be derived from it. The use of a classical metalanguage does not prejudge in any way the hypotheses about the colloquial *if–then* of the object language which it is used to describe. All that is assumed in the choice of a classical metalanguage is that a classical metalanguage is rich enough to express the hypotheses that will be of interest. 'If–then' is associated with the material conditional in the metalanguage only as a 'convenient oral reading' as Church put it, with no claims made about how good a reading it is.

8.8 Exact Statement of Hypothesis B

For the simple sentences the credibility function and learning operation are determined by the general definition of a semantically structuralizable probabilistically weighted language: the credibility of a simple sentence is the sum of the current probabilities associated with the modelling structures which satisfy it, and the operation of learning a simple sentence involves raising these probabilities in a manner consistent with the Dynamic Principle of Indifference. Hypothesis B is essentially a set of rules extending this credibility function and learning operation from simple to compound sentences. Hypothesis B is unlike Hypothesis A in that it does *not* assume that there is a way of extending the satisfaction relation to the compound sentences. Hypothesis B is instead stated directly in terms of the credibility function and learning operation without any appeal to an underlying satisfaction relation.

The rules which constitute Hypothesis B can be stated in a preliminary informal way as follows:

B1. The credibility of a sentence of the form *if* s_1 *then* s_2 in an information state is equal to the credibility of s_2 in the new state which would result from the original state if s_1 were learned in it with high credibility.

B2. The credibility of s_1 *and* s_2 in an information state is equal to the credibility of s_1 in that state times the credibility of s_2 in the new state which would result if s_1 were learned with high credibility.

B3. The credibility of s_1 *or* s_2 in an information state is equal to the credibility of s_1 in that state plus the product of one minus that credibility and the credibility of s_2 in the new state which would result if s_1 were learned with low credibility (i.e. 'unlearned').

B4. The credibility of *not-s* in an information state is equal to one minus the credibility of s in that state.

B5. The learning operation conforms to the Dynamic Principle of Indifference.

When formulated mathematically, these rules impose a set of constraints on credibility and learning which, according to Hypothesis B, are equivalent to the constraints imposed by the rules of English.

Before these rules can be stated with mathematical precision it is necessary to generalize the notion of a DPI-transform (p. 145). The generalization is from DPI-transforms originating in one sentence to DPI-transforms originating in any finite number of sentences. Suppose $a_1, \ldots, a_n$ are sentences in the sentence set of a semantically structured probabilistically weighted language automaton with model set M and satisfaction relation H. Consider the family (field) of subsets of M generated from $H[a_1], \ldots, H[a_n]$ by the operations of union, intersection, and complementation with respect to M. The non-empty subsets in this family which are minimal in the sense that they properly include no other non-empty members of the family are its *atoms*. For any pair of states z and z' in the automaton, we will say that z' is a *DPI-transform* of z *originating in* $a_1, \ldots, a_n$ if and only if for every atom M' in the family of subsets of M generated by $H[a_1], \ldots, H[a_n]$ there exists a non-negative real number α such that for every modelling structure m in M', $z'(m) = \alpha z(m)$. The idea of the definition is simply that the ratios of the probabilities of the elementary events within an event are maintained by a DPI-transform provided no sentence involved in the transform is satisfied by some of those elementary events but not others.

Now consider a semantically structured structuralization of the sublanguage of English consisting of the simple sentences. Since by assumption it has a model set and a satisfaction relation it is meaningful to speak of DPI-transforms in it originating in sets of simple sentences. Under Hypothesis B we keep this model set and satisfaction relation and extend the sentence set of the structuralization to include all the sentences in S, making the following constraints on the credibility function C and learning operation L of the extended structuralization:

B1. $C(z, \textit{if } s_1 \textit{ then } s_2) = C(L(z, s_1, 1), s_2);$

B2. $C(z, s_1 \textit{ and } s_2) = C(z, s_1)C(L(z, s_1, 1), s_2)$

B3. $C(z, s_1 \textit{ or } s_2) = C(z, s_1) + (1 - C(z, s_1))C(L(z, s_1, 0), s_2)$

B4. $C(z, \textit{not-}s) = 1 - C(z, s)$

B5. For every state z, sentence s, and number ρ between 0 and 1, $L(z, s, \rho)$ is a DPI-transform of z originating in the set of all simple sentences which occur in s.

Hypothesis B is the claim that the sublanguage of English whose sentence set is S has a structuralization $\langle S, W, Z, L, C\rangle$ which is stable and receptive (p. 134) and constrained by B1-B5.

The expressions of the form '$L(z, s, 1)$' and '$L(z, s, 0)$' which appear in B1-B3 require some explanation in view of the assumption that no non-tautologous sentence can ever attain a credibility of exactly 1 or exactly 0 (Section 7.3). The intended interpretation of these expressions is that when necessary they are to be regarded as abbreviations of longer statements involving limit operations. If for example s_1 does not happen to be a logically valid sentence, then the right side of equation B1 is to be understood as a shorthand way of writing

$$\lim_{\rho \to 1} C(L(z, s_1, \rho), s_2).$$

Similar remarks apply to the problematic credibility expressions in B2 and B3.

A minor technical point concerning the interpretation of B2 and B3 is that an undefined quantity multiplied by zero is to be assumed to be zero, not undefined. In B2, for example, if s_1 happens to be tautologously false then

$$\lim_{\rho \to 1} C(L(z, s_1, \rho), s_2)$$

is undefined. But since the other credibility expression by which it is multiplied is zero under these circumstances the credibility of the conjunction as a whole is zero.

8.9 Remarks on Hypothesis B

1. In the special case of conditionals *If a then b* in which *a* and *b* are simple sentences, B1 has the effect of postulating that the credibility weight is a conditional probability. That is, B1 states in effect

$$C(z, \textit{If } a \textit{ then } b) = z(H[a] \cap H[b])/z(H[a]).$$

This is because learning *a* to the credibility level 1 is the dynamic counterpart of the probabilistic notion of conditionalizing on *a*.

2. In particular, B1 leaves the credibility of *If a then b* undefined when the credibility of *a* is zero just as conditional probabilities are undefined when the event conditionalized upon has zero probability. Thus according to Hypothesis B the rules of English involve an element of 'linguistic incompleteness' (cf. Section 9.4).

3. Since the credibility function is never undefined in a semantically structuralizable language, the element of incompleteness alone shows that Hypothesis B cannot be reconciled with the idea that English is semantically structuralizable in the 'narrow' sense (Section 7.8). Rather, Hypothesis B is an example of the use of 'extended' semantics.

4. By way of motivation of B2, it may be helpful to note its analogy with the following identity of the probability calculus when only simple sentences are involved:

$$Pr(E_1 \cap E_2) = Pr(E_1)Pr(E_2/E_1)$$

where E_1 and E_2 are events in an event space. Similarly B3 is reminiscent of the identity

$$Pr(E_1 \cup E_2) = Pr(E_1) + (1 - Pr(E_1))Pr(E_2/\sim E_1),$$

and B4 is suggestive of

$$Pr(\sim E) = 1 - Pr(E).$$

5. When the connectives are used to join together sentences which themselves contain *if–then*, all simple-minded analogies with the standard probability calculus seem to fail, and it is probably simplest to forget classical probability theory and interpret B1-B4 directly in terms of the dynamic learning process.

6. The fact that there is no easy translation of B1-B4 into the standard probability calculus for sentences whose components contain *if–then* is reflected in the interpretation of the credibility weights. The credibility weights of simple sentences are interpretable as straightforward personal probabilities; their behavioral meaning is explainable in terms of betting propensities as discussed earlier (Section 7.2). Sentences of form *If a then b* with a and b simple are interpretable as conditional probabilities for which it is possible to give a behavioral meaning in terms of bets that are called off under certain circumstances. But the credibility weights of more complicated sentences have in general no straightforward interpretation as probabilities of an ordinary kind. Their primitive behavioral interpretation is that of acceptability or assertability signals passing between the Language Automaton and other behavioral entities. Beyond this they may have no interpretation beyond that lent them by B1-B5 themselves, which is to say their interpretation as sometimes rather complicated dynamic functions of the probabilities of the simple sentences they contain.

7. The main constraint on the learning operation L is the assumption of receptiveness. It shifts the burden of the detailed statement of Hypothesis B with regard to L back onto postulates B1-B4 about C by requiring of L that after the learning has taken place the resulting new state will be consistent with the constraints on C.

8. B5 states the only other general constraint on L. It is not a sufficiently tight constraint to determine uniquely (together with B1-B4 and receptiveness) what the results of learning will be in all cases. Hypothesis B does not define L and C categorically; it states many but not all of their properties.

9. This doesn't mean, though, that Hypothesis B is unfinished in the sense that one ought to look for more logico-linguistic postulates to add to B1-B5. B1-B5 stop short of uniquely determining L and C, but that is because of the linguistic incompleteness of the rules of English. Or so Hypothesis B claims.

10. That Hypothesis B does not determine L uniquely is well illustrated by the learning of conditionals with simple antecedents and consequents. According to B5, when the learning of *If a then b* takes place all subevents of the event $H[a] \cap H[b]$ must rise or fall by the same proportion, and the same is true within $H[a] \cap {\sim}H[b]$, ${\sim}H[a] \cap H[b]$, and ${\sim}H[a] \cap {\sim}H[b]$. But apart from this B5 and B1 require only that after the learning takes place the ratio of $H[a] \cap H[b]$ to $H[a]$ must be at the specified figure, and there is a degree of freedom left over. To illustrate, suppose before receiving *If a then b* the information state is that of Figure 8.1(a). Figures 8.1(b-d) show three possible results of learning *If a then b* to a credibility level of 0.75. There is no way to predict *on linguistic grounds alone* which of the three if any will be the new state, for the probability of b given a is 0.75 in all three.

11. Nevertheless, even though it may be unpredictable on linguistic grounds alone, it may be possible to predict the new state on combined linguistic and nonlinguistic grounds. Figure 8.2(a) is identical with 8.1(a) except that the event space has been expanded to allow the representation of a new event associated with the possible *utterance* of *If a then b* in the hearer's presence. The new event is associated with a token rather than a type, and so is not the kind of event with which ordinary linguistic rules are expected to deal. But conceptually one can see that if the utterance occurs and its probability rises to nearly 1 in the hearer's mind, by invoking DPI one can predict that the new state will be the one in Figure 8.2(b), which is essentially the same as 8.1(d).

12. It might be thought that stronger linguistic constraints could be stated for the learning of conditionals, but a consideration of actual English examples shows this to be problematical. Suppose for example someone were to suggest the additional constraint that when *If a then b* is learned, the credibility of the antecedent, a, remains unchanged. To test this idea,

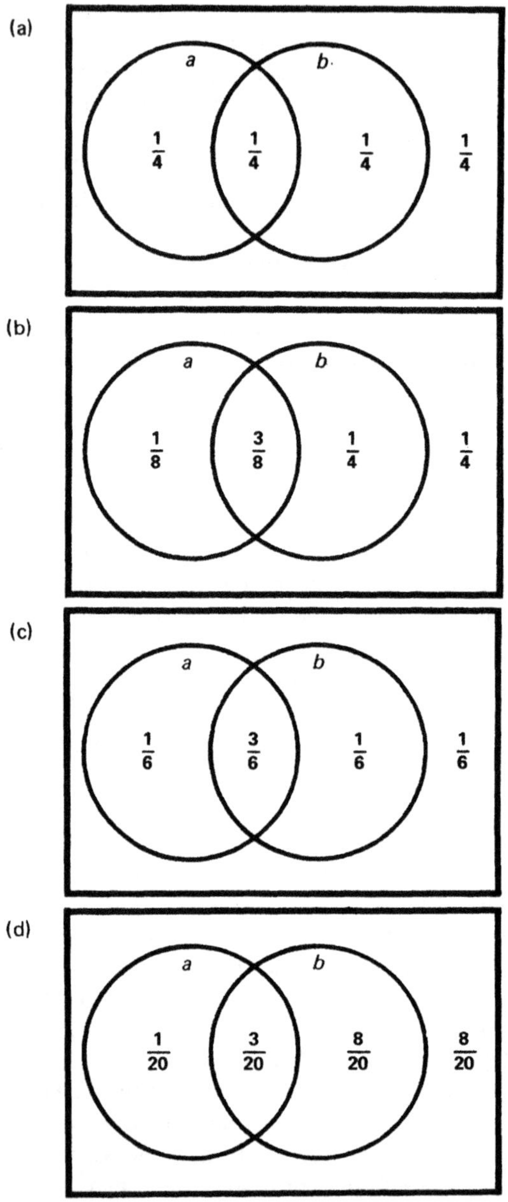

Figure 8.1. Venn diagrams showing three possible results of learning *If a then b* in such a way as to raise the corresponding conditional probability from 0.5 to 0.75. If (a) is the probability distribution prior to receiving the message *If a then b*, will the posterior state be that of (b), (c), (d), or none of these?

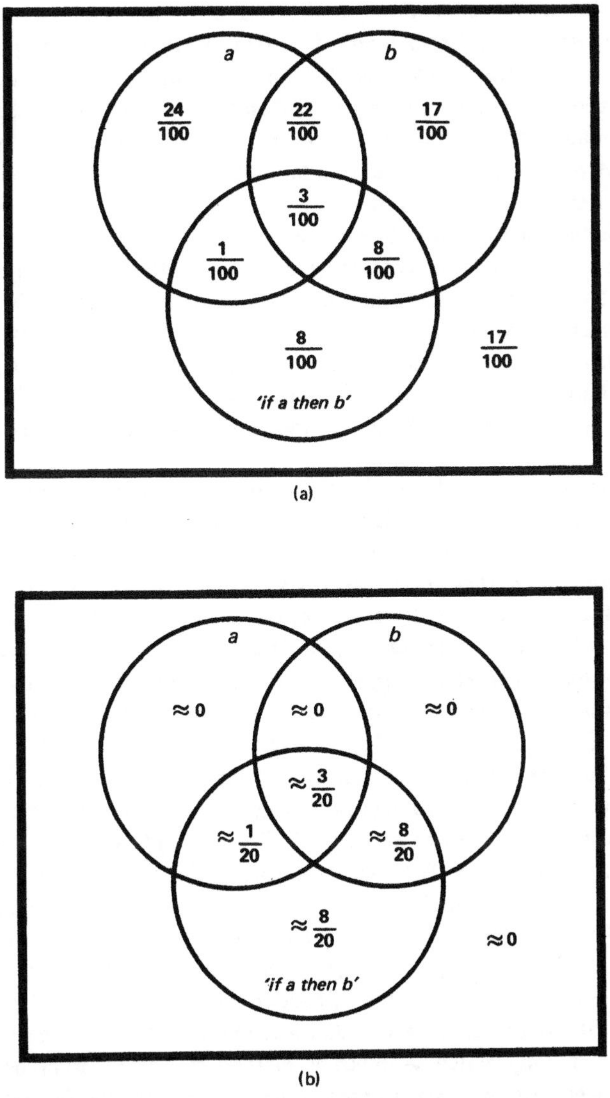

Figure 8.2. Venn diagram of an information state before and after receiving the message *If a then b*. The indistinguishability classes representing possible states of affairs have been subdivided in such a way as to allow representation of the event of hearing an utterance of *If a then b*. When the utterance is heard the personal probability of this event rises from 0.2 to a value close to 1.

consider the sentence *If these scales are accurate then you weigh four hundred pounds*, said to a man of normal weight who has just weighed himself on a weighing machine, by his trustworthy companion who has just read the ticket that popped out. Common sense tells us that the man will assimilate his companion's statement mainly by drastically reducing his subjective credibility for the statement *These scales are accurate*, contrary to the suggested constraint.

13. Some might respond to this example by proposing instead the weaker constraint that at least the crediblity of the antecedent never gets *larger* when a conditional sentence is learned. But even this suggestion becomes suspect when we change the example to read *If these scales are accurate then you weigh exactly one hundred and fifty pounds*, said to a man who happens to know already that he weighs exactly one hundred and fifty pounds. His opinion of the likelihood that the scales are accurate is apt to go *up*.

14. Finally one might be tempted to say that if no other additional constraints can be found, at least it is obvious that the credibilities within the event $\sim H[a]$ rise or fall uniformly when *If a then b* is learned, since the statement *If a then b* simply doesn't make any assertion about what happens if a isn't so. But there are apparent counterexamples even for this. Imagine a sports announcer saying, while awaiting the judges' final decision, *Well, if he lost, he certainly tried his hardest*. One would normally assume from this that even if the contestant won he tried his hardest, i.e. the learner's ratio of $\sim H[a] \cap H[b]$ to $\sim H[a] \cap \sim H[b]$ increases.

15. The phenomenon of learning indeterminacy is not confined to conditional statements; it infects sentences of the form *a and b* too. Consider the sentence *Switch X is on and switch Y is on*, said by a man who has been feeling around for both switches in the dark. On hearing this utterance his companion would normally, in addition to raising his probability for $H[a] \cap H[b]$ in accordance with B2, increase somewhat his ratio of $H[a] \cap \sim H[b]$ to $\sim H[a] \cap \sim H[b]$, and similarly of $\sim H[a] \cap H[b]$ to $\sim H[a] \cap \sim H[b]$, reasoning that although the man might well be mistaken about one of the switches he is much less likely to be mistaken about both. Thus when a conjunctive sentence is learned the event complementary to the joint event of principal interest does not necessarily fall uniformly in the learner's subjective probability distribution. However, on occasion it does; consider e.g. the same sentence uttered on the grounds, surmised by the hearer, that the speaker has discovered that an appliance is on to which both switches are known to be wired in series.

16. We see then that under Hypothesis B the learning of a conjunctive sentence of form *a and b* does not necessarily have the same result as the learning of *a* followed at a later time by the learning of *b*, though it could have. That there can be a difference between the learning of a conjunctive statement and the consecutive learning of its component statements is strikingly illustrated by the case in which the components are logically inconsistent. Here the results cannot be the same. If a man were to utter sentence of form *a and not-a* his hearer would classify it as self-contradictory, or at any rate accord it special treatment. If on the other hand the man utters *a*, leaves the room for a while, comes back, and utters *not-a*, nothing special happens; normally his hearer will first give *a* a high probability, then lower it again later upon hearing the man's second utterance.

17. What Hypothesis B says about English disjunctions is analogous. When a sentence of form *a or b* is learned, the credibilities of the events $H[a] \cap H[b]$, $H[a] \cap \sim H[b]$, and $\sim H[a] \cap H[b]$ do not necessarily all rise uniformly. As a case in point imagine a detective who has just discovered some fresh evidence saying to his partner, *I still don't know any better than we did before whether our suspect is guilty or not, but I am now convinced of this: either he is guilty or he was very cleverly framed.* His partner is apt on hearing this disjunctive assertion to raise his personal probability for the proposition that the suspect was framed, which we may suppose is a new idea to him, by a much higher ratio than the ratio by which he raises his probability for the proposition that the suspect is guilty.

18. Confining attention for a moment to conditional-free sentences, what Hypothesis A (with A1 deleted) says about English *and*, *or*, and *not*, is consistent with but stronger than what Hypothesis B (with B1 deleted) says about them. This is because Hypothesis B leaves, as discussed, a certain degree of freedom in the specification of the learning operation while Hypothesis A does not.

19. This difference notwithstanding, it can be shown that the deductive *logic* of the conditional-free statements is the same under Hypothesis B as under Hypothesis A; i.e. it is just the familiar classical logic of conjunction, disjunction, and negation. This is an illustration of earlier remarks to the effect that languages are not entirely determined by their logics, though they may be largely so.

20. When the restriction to conditional-free sentences is removed, however, the logic obtained under Hypothesis B differs strikingly from the logic obtained under A, as we will now begin to see.

8.10 CONTRAPOSITION

It is time to start looking seriously at some of the logico-linguistic evidence for and against the two hypotheses. As a start let us consider more closely the sports announcer's statement, *Well, if he didn't win, he certainly tried his hardest*, uttered while awaiting the judges' decision. Would the announcer who said this necessarily have to believe as well that *If he didn't try his hardest, he won*? Obviously not. We may therefore record as a bit of evidence about English usage that

If he didn't win, he tried his hardest.

* ———————————————

If he didn't try his hardest, he won.

The asterisk indicates that this argument is *not* a valid logical implication in English, our sports announcer being a counterexample to the claim that any rational English speaker who believes the first sentence must also believe the second.

Now let us see what Hypothesis A predicts about this argument. According to Hypothesis A any argument from *If not-a then b* to *If not-b then a* is valid; it should be impossible for any rational person to have at the same time a high credibility level for the first statement and a low credibility level for the second. In fact, this argument is an instance of the well-known Principle of Contraposition, according to which any implicative statement logically implies its contrapositive. The Principle of Contraposition is readily provable in a number of forms from A1-A4, or alternatively (using the fact that Hypothesis A yields the classical logic) by more traditional arguments involving truth tables or classical axiom systems. Hypothesis A therefore predicts in particular that the above English argument is valid, i.e. Hypothesis A appears to be in conflict with this piece of evidence.

Hypothesis B, on the other hand, does *not* predict that *If not-a then b* must have *If not-b then a* among its logical consequences. Under Hypothesis B it is possible for a rational person to believe the first of these conditional statements at a high credibility level and the second one at a low level. This is possible in case the language user's subjective probability distribution is as shown in Figure 8.3, where the size of the area within an event is to be understood as indicating its relative probability. In that figure one sees that the conditional probability of *b* given *not-a* is large while the conditional probability of *a* given *not-b* is small. Moreover, by stretching suitable sectors of the diagram one could make the first of these conditional probabilities

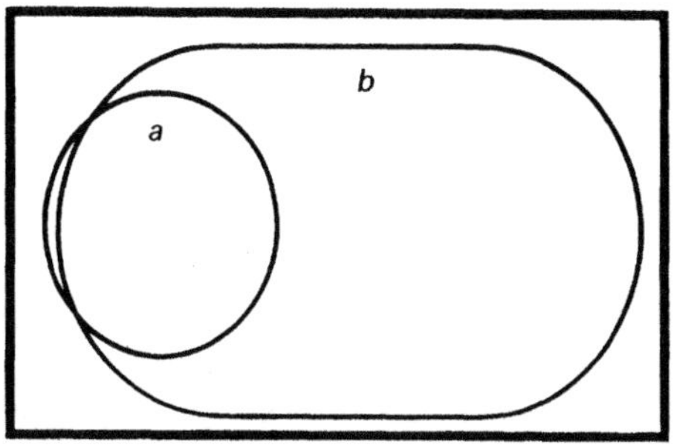

Figure 8.3. Venn diagram of an information state in which, under Hypothesis B, *If not-a then b* has a high credibility weight but *If not-b then a* a low credibility weight.

come *arbitrarily* close to 1 while the second becomes no higher. Hence by our criterion (Section 7.3) the logical implication fails.

The probability of *b* itself has to be large for this situation to exist, and one senses that in the present example it *is* large: the announcer is quite convinced that the contestant tried his hardest. In fact, Hypothesis B could have been used to predict that our sports announcer would *not* necessarily believe the conclusion of this specious argument. On this first bit of evidence, then, Hypothesis B has come off better than Hypothesis A.

To obtain some feeling for whether this evidence is just an isolated stylistic quirk of English or an example of a more general phenomenon, it is helpful to see if other examples can be found that exhibit the same properties. As it happens, at least two similar examples have been mentioned already in the literature. They are

If it rains, there will not be a tremendous cloudburst.
* ―――――――――――――――
If there is a tremendous cloudburst it will not rain.

and

There's some sugar in the cupboard if you want some.
* ―――――――――――――――
If there isn't any sugar in the cupboard you don't want any.

More examples are easily generated; in fact, any situation involving a belief state of the sort illustrated in Figure 8.3 seems to provide English counter-examples to the law of contraposition. We conclude tentatively that it was not just something special about the subject matter of the sports announcer example that made it appear to violate the principle of contraposition. The rule of contraposition simply fails for those English examples for which Hypothesis B predicts it will fail.

Hypothesis B has won the first round.

8.11 Methodological Review

It may be helpful to go over in slow motion what just took place, for if those transactions can be understood clearly enough a paradigm will have been established for the examination of all future evidence. Our underlying scientific methodology is the hypothetico-deductive method (Section 3.3). It involves drawing out from the hypothesis to be tested particular logical consequences in the form of basic evidence statements (Section 3.4) and comparing these consequences against observational evidence obtained by informant technique. If a logical consequence of the hypothesis turns out to be in agreement with the evidence, the hypothesis becomes slightly more probable in our minds; if it conflicts with the evidence we should consider rejecting it.

It is permissible to begin either by first finding some evidence and then drawing out the relevant consequence of the hypothesis or *vice versa*. Let us start by gathering the evidence. First we satisfy ourselves that both

If he didn't win, he tried his hardest

and

If he didn't try his hardest, he won

are well-formed sentences. This we can do by using ourselves as informants and asking ourselves whether the sentences could be uttered by native English speakers under conditions in which the utterance would be thought 'syntactically correct', or 'grammatically proper'. Since this sort of informant technique has been the subject of much discussion in the linguistic literature, we pass over it quickly here. We must also check whether the sentences are indeed in the sublanguage S of English in which we are interested. Since *He didn't win* and *He tried his hardest* are declarative and seem free of at least the more obviously problematic connectives, modal auxiliaries, etc., we assume tentatively that they are in the semantically structuralizable sub-language of English. Since these two simple sentences are joined by the *if–then* connective in its *then*-less variant, *If he didn't win he tried his hardest* does seem to be a member of the set of sentences S we are interested in, and so does *If he didn't try his hardest he won.*

The next step is to check whether these sentences have any ambiguities, hidden variables, or other problems of interpretation lurking in them. There do not seem to be any serious syntactic ambiguities so no brackets need be inserted. There is an indexical expression, the *he*, but to get around it the informant need only be instructed to assume that the *he* (and *his*) refer to the same man in both sentences. The temporal points of reference of the two sentences may likewise be assumed the same. There are no obvious semantic ambiguities and the hidden variable problem does not seem serious. We conclude that if there are interpretive problems, they are fairly subtle.

The ground is now prepared for the crucial question of whether the first sentence logically implies the second in English. Does the evidence indicate that the question mark should be replaced by an asterisk or a blank in the following argument?

If he didn't win, he tried his hardest.
? ———————————————
If he didn't try his hardest, he won.

According to the definition of logical implication (Section 7.3) the basic evidence statement to be examined is, paraphrased informally, the meta-statement

> By selecting information states in which the credibility weight of *If he didn't win, he tried his hardest* is sufficiently high, the credibility level of *If he didn't try his hardest, he won* can be forced to be arbitrarily high.

We must ask ourselves as informants whether we can imagine rational belief states in which the credibility weight of the first sentence approaches 1 but not the second. Upon performing the required gedanken experimentation we find we can – in fact our hypothetical sports announcer is in such a belief state. (Speaking more cautiously I should say that *I* can imagine a rational person in such a belief state; if the reader cannot it may be that his dialect differs from mine.) The basic evidence statement in question therefore runs counter to the evidence gathered by informant technique and we may record

If he didn't win, he tried his hardest.
* ———————————————
If he didn't try his hardest, he won.

The slightly more elaborate informant technique of considering What-Do-You-Know? games could also be used to gather this bit of evidence. The procedure would have been to show the informant a game-protocol such as the following one.

Q: DO YOU BELIEVE AT A CREDIBILITY LEVEL OF 0.9 THAT *if he didn't win, he tried his hardest*?
A: Yes.

Q: DO YOU BELIEVE AT A CREDIBILITY LEVEL OF 0.3 THAT *if he didn't try his hardest, he won*?
A: Yes.

Here the figures 0.9 and 0.3 are interpretable as conditional probabilities and can be explained to the informant in behavioral terms using the idea of a conditional bet. The informant is to study this game to see if he can conceive of a perfect Answerer so informed initially that had he been asked the questions in the game he would have answered as indicated. An alert informant should see that nothing prevents a rational English speaker, say a sports announcer, from being so informed sometimes; and that moreover nothing would prevent it even if the figure 0.9 were replaced by numbers closer and closer to 1. By this informant technique too, then, the basic evidence statement of interest is falsified. The What-Do-You-Know? game technique is of course only an elaboration of the more casual method of interpreting the basic evidence statement directly.

Having recorded this evidence the next step is to see whether it is consistent with the observational consequences deducible from the hypotheses of interest. From Hypothesis A (using rules A1, A4 and Theorem 7.2) it can be deduced that every state in which a sentence of form *If not-a then b* has sufficiently high credibility is also one in which *If not-b then a* has high credibility, i.e. the basic evidence statement displayed earlier is mathematically derivable from Hypothesis A. From Hypothesis B the *negation* of this basic evidence statement is derivable by exhibiting any sequence of states in which the measure over $\sim H[a] \cap H[b]$ divided by the measure over $\sim H[a]$ approaches 1 as a limit while the measure over $H[a] \cap \sim H[b]$ divided by the measure over $\sim H[b]$ stays constant or declines. Since it is the negation of the basic evidence statement that was found to concur with observation, Hypothesis B is consistent with the evidence and Hypothesis A is not.

If one were to follow the hypothetico-deductive method blindly, one would have to throw out Hypothesis A at this point. However it is wisest not to abandon Hypothesis A entirely just yet. There are many uncertainties about this or any one particular bit of linguistic evidence. Perhaps the allegedly simple sentences involved are not really in the semantically structuralizable part of English; perhaps there are hidden variables in other interpretational problems that were not unearthed; perhaps the particular usage of *if–then* examined as evidence is just a stylistic oddity unrelated to normal usage. One who is willing to defend a hypothesis sufficiently stubbornly can usually think of *ad hoc* explanations for almost any piece of apparent counterevidence in isolation. Abandonment of a hypothesis should and does come only when the *ad hoc* explanations required to sustain it start to pile up to the point of absurdity, while some alternative hypothesis requires fewer or no *ad hoc* explanations.

The entire process may be repeated for other English arguments of the form

$$? \frac{\textit{If not-a then b}}{\textit{If not-b then a}} .$$

If the outcome is similar in spite of considerable variation in subject matter and stylistic detail, one gains confidence that one was not led astray in the first instance by mere problems of interpretation or hidden variables. Note though that an argument of this form that seems valid is not necessarily counterevidence against Hypothesis B. Even according to Hypothesis B such arguments often sound valid in the sense that people who believe the premise would *usually* believe the conclusion. It is only whose who believe the premise in an information state with the special characteristics of Figure 8.3 who fail to believe the conclusion, according to Hypothesis B.

8.12 THE HYPOTHETICAL SYLLOGISM

Resuming speed, we examine next the argument

If Brown wins the election, Smith will retire to private life.
If Smith dies before the election, Brown will win it.

* ―――――――――――――――――――――――――――――――

If Smith dies before the election, then he will retire to private life.

This putative logical implication is clearly invalid by our definitions, for it would be quite possible for an English speaker to believe strongly in the premises while disbelieving the conclusion. In fact, disbelieving this conclusion would be the normal attitude for a rational person acquainted with the facts of death and retirement. Now under Hypothesis A the argument is valid; it is in fact an instance of the well-known 'hypothetical syllogism' of classical logic which allows one to deduce *If a then c* from the premises

If b then c and *If a then b*. But under Hypothesis B it is not valid, so once again we have a piece of evidence that contradicts Hypothesis A but supports Hypothesis B.

The reason why it is possible under Hypothesis B for the premises of a hypothetical syllogism to have high credibility while the conclusion has low credibility is that a speaker's subjective probability distribution might be as shown in Figure 8.4. There the conditional probability of *c* given *a* is low in spite of the fact that the conditional probabilities of *b* given *a* and *c* given *b* are high. For this situation to arise *a* must not be too large and it must have little chance of happening jointly with *c*. And that accurately describes the above argument: that Smith will die before the election is not highly probable, and it is improbable in the extreme that Smith will die and then retire. Thus once again Hypothesis B has successfuly predicted the invalidity of the argument form.

With Figure 8.4 as a guide it is possible to construct any number of similar examples. Consider for instance the argument

If I am ever in California I shall take up yachting.
If I commit a crime and am sentenced to San Quentin Prison then I shall be in California.

* ———————————————

If I commit a crime and am sentenced to San Quentin Prison then I shall take up yachting.

This sounds suspicious to say the least, so we have another counterexample to the Hypothetical Syllogism in English.

Of course, not all English instances of the hypothetical syllogism sound immediately suspicious. Offhand it might sound plausible to make the inference

If I am ever in California I shall take up yachting.
If I move to the San Francisco area I shall be in California.

* ———————————————

If I move to the San Francisco area I shall take up yachting.

But this inference sounds valid at first only because most people who assert the premises would usually *happen to* believe the conclusion at a high credibility level as well. To see that not all people would necessarily so believe it, consider the case of a San Francisco-hater who thinks that he might well move to the Los Angeles area, but who thinks that the only real possibility of hiş moving to the San Francisco area would be the extremely remote one of his committing a crime and being sentenced to San Quentin Prison. A person in such a belief state could well be as much a counterinstance to

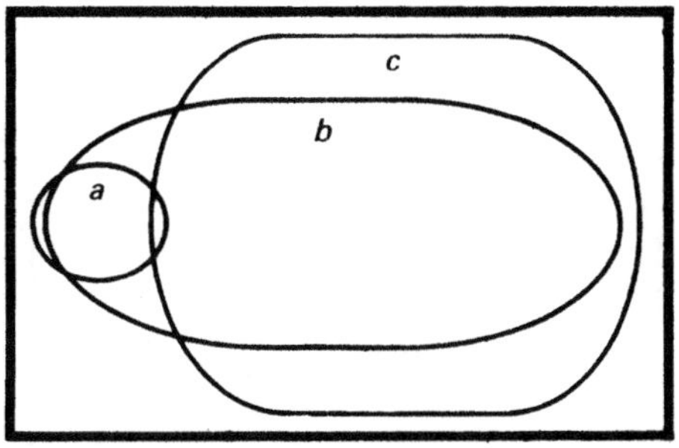

Figure 8.4. An information state in which, under Hypothesis B, *If a then b* and *If b then c* both have high credibility weights but *If a then c* has a low credibility weight.

the second argument as to the first. This shows how careful an informant must be to think through all the possibilities before declaring a putative logical implication relationship to be valid. (There are ways by which a careful informant could make sure he has considered all the important cases, e.g. schemes involving the systematic exploration of Venn diagrams of all possible sorts as a way of suggesting different possible information states.)

8.13 FURTHER INFERENCE PATTERNS

Under the classical logic of Hypothesis A, a sentence *If a then b* logically implies *If a and c then b* for any *a, b,* and *c*. But consider

If there is good weather tomorrow, the game will be played.

* ———————————————————————

If there is good weather tomorrow and all of the players fall ill, the game will be played.

In English this argument is surely invalid, for no matter how strongly someone were to believe that the game will be played, weather permitting, it would always be possible that he believed even more strongly that a completely unexpected illness of all the players would endanger the game. Under Hypothesis B the implication fails accordingly, because of the possible existence of a subjective probability distribution of the kind shown in Figure 8.5.

Under Hypothesis A (and classically), *If b then not-c* combined with *If a then c* logically implies *If a then not-b*. But consider the following

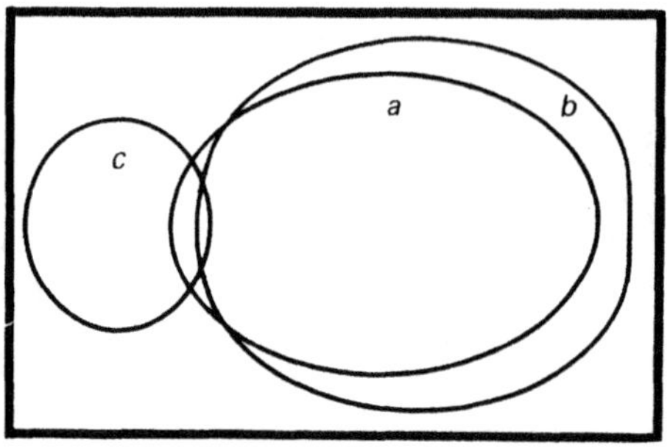

Figure 8.5. An information state in which *If a then b* has high credibility while *If [a and c] then b* has low credibility.

argument about Jones, an impecunious employee enrolled in a pension plan which does not allow for early retirement:

> *If Jones retires before he is sixty-five, he won't be financially secure.*
>
> *If Jones unexpectedly receives a large inheritance tomorrow, he will be financially secure.*
>
> * ———
>
> *If Jones unexpectedly receives a large inheritance tomorrow, then he will not retire before he is sixty-five.*

Under Hypothesis B this implication fails because of the possibility of probability distributions such as the one in Figure 8.6, which plausibly represents what one tends to feel about Jones.

Classically, *if* [*a and b*] *then c* logically implies *if* [*a and not-c*] *then not-b*. An English example is

> *If you strike this match and wear a hat, the match will light.*
>
> * ———
>
> *If you strike this match and it doesn't light, then you won't be wearing a hat.*

Though endorsed by Hypothesis A this argument is not valid under Hypothesis B; the construction of the Venn diagram showing why is left to the reader.

Things are looking bad for Hypothesis A.

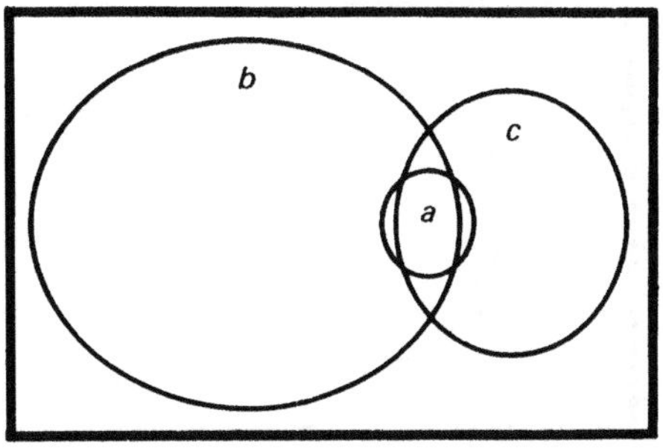

Figure 8.6. An information state in which the credibility weights of *If b then not-c* and of *If a then c* are high but the credibility of *If a then not-b* is low.

8.14 THE PARADOXES OF MATERIAL IMPLICATION

The so-called Paradoxes of Material Implication have been variously stated. One way of stating them is

PARADOX I: *Not-a* logically implies *If a then b.*

PARADOX II: *b* logically implies *If a then b.*

Viewed as assertions about English *if–then* (and this is the only interpretation of the paradoxes that is germane here), these statements can be tested out as logico-linguistic hypotheses.

In connection with Paradox I, consider the argument

John will not arrive on the ten o'clock plane

* *If John does arrive on the ten o'clock plane, he will have had lunch on board.*

This inference is valid classically and under Hypothesis A, but it is counter-intuitive and is empirically invalid under our informant technique. No matter how close to being certain an English speaker might be of the premise, he might still be unsure of the conclusion. The argument is not valid under Hypothesis B, as Figure 8.7 shows. Thus what has traditionally been regarded by many as a philosophical paradox, we may regard simply as logico-linguistic evidence against Hypothesis A and for Hypothesis B.

It is fair to ask, "If the situation is so simple, why was this 'paradox' ever regarded as paradoxical?" Possibly part of the reason is that it has not always been recognized clearly that issues discussed under the heading of

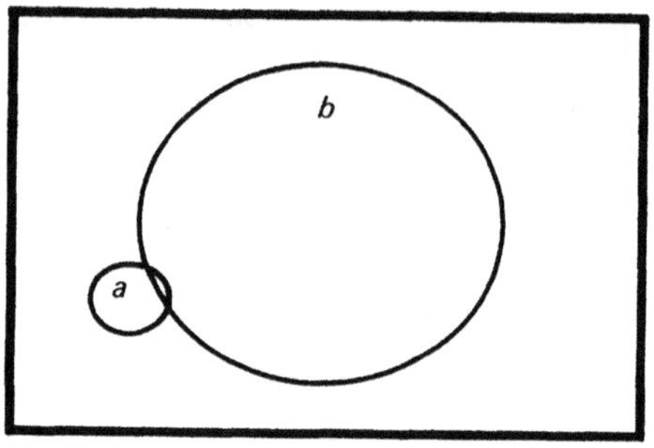

Figure 8.7. An information state in which the credibility level of *Not-a* is high but that of *If a then b* low.

'philosophical logic' have often been logico-linguistic in nature. As soon as one adopts the hypothetico-deductive framework, the double nature of a statement such as "'*not-a*' logically implies '*If a then b*'" becomes clear: the statement is true in the sense that it is entailed by the classical propositional calculus considered as a hypothesis, but false in the sense that this hypothesis conflicts with the linguistic evidence about *if–then*. The situation is no more mysterious than any other instance in which what a theory predicts turns out to conflict with observation.

For Paradox II the situation is similar. Hypothesis A predicts the validity of the counterintuitive argument

John will arrive on the ten o'clock plane

* ———————————————

John will arrive on the ten o'clock plane if he missed it in New York.

But this is not a valid logical argument under Hypothesis B, as Figure 8.8 shows. Hence the second paradox too has for us only the force of a bit of evidence against Hypothesis A and for Hypothesis B. However, the second paradox does involve an additional complication that the first does not, as will be seen next.

8.15 THE SECOND PARADOX RE-EXAMINED DYNAMICALLY

When one thinks in terms of the process of growing surer of a statement, there is something unsatisfying about the piece of linguistic evidence just cited in connection with Paradox II. Imagine John's wife waiting for John in the airport and continually receiving more and more reassurance of the truth

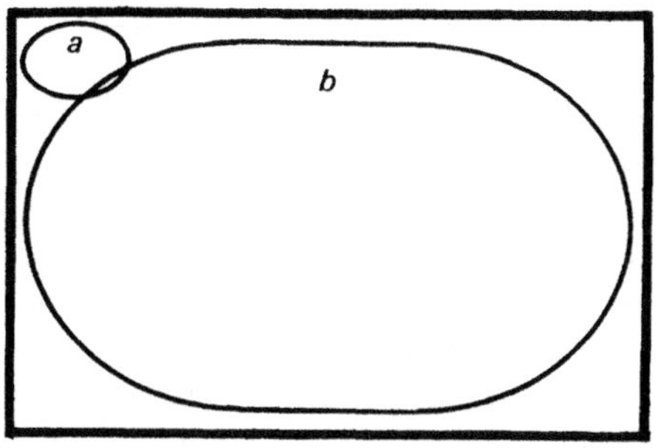

Figure 8.8. An information state in which the credibility level of b is high while that of *If a then b* is low.

of the statement that he will arrive on the ten o'clock plane: first she receives a succession of telegrams saying so, then the pilot radios her that John is on the plane, then as the plane lands a friend with field glasses says he is on it, etc. Sooner or later, one feels, her growing certainty that John will arrive on the plane would overwhelm all other considerations, and the wife would eventually come to assent even to the statement that *John will arrive on the ten o'clock plane if he missed it in New York.* If John missed this plane in New York, he must have done something like taking a faster one and intercepting this one at Chicago, she might reason. For any given *if*-clause, no matter how improbable it is initially, she would eventually grow certain enough of John's arrival to believe the conditional statement stating he will arrive, conditionally on that *if*-clause.

The situation may be described technically by saying that even though the argument is not valid logically, it is nevertheless valid dynamically (Section 7.3). That is, our informant procedure leads us to record

John will arrive on the ten o'clock plane
―――――――――――――――――――――― ↓
John will arrive on the ten o'clock plane if he missed it in New York.

as a valid instance of dynamic implication. If a rational person were to grow surer and surer of the premise (in a manner consistent with DPI) then he would eventually have to grow sure of the conclusion too, and this seems to hold true of other English examples of the same inference pattern so long as the two simple sentences involved are not logically inconsistent.

That *b* dynamically implies *If a then b* for logically compatible *a* and *b* is successfully predicted by both Hypotheses A and B. That Hypothesis B predicts it might seem surprising insofar as the implication is not logically valid under B. But the conflict between logical and dynamic implication under Hypothesis B can be understood by studying Figure 8.8 once again. Logical implication fails because by making the area within *b* but outside of *a* sufficiently large, the credibility of *b* can be made arbitrarily high while that of *If a then b* remains low. But it can also be seen that when *all* of the area within *b* is stretched uniformly, as growing surer and surer of statement *b* requires, *If a then b* must also attain a higher and higher credibility.

The difference between logical and dynamic implication is a subtle one which only a precise theory of pragmatics could hope to bring out. The distinction is more than academic, though, as Paradox II shows. It may be that inattention to the distinction has been partly to blame for the confusion surrounding this paradox.

8.16 MODUS PONENS AND MODUS TOLLENS

We have just seen that under Hypothesis B an instance of implication can fail logically and yet be valid dynamically. The reverse situation can also occur: under Hypothesis B an implication can hold logically and fail dynamically. An inference pattern which illustrates this particularly dramatically is none other than the famous 'Modus Ponens' or 'Rule of Detachment' which has historically been so pivotal in many traditional axiomatizations of the classical logic.

Returning to an earlier example we may record

These scales are accurate
If these scales are accurate then John weighs four hundred pounds.
* ―――――――――――――――――――――――――― ↓
John weighs four hundred pounds.

To see that this argument is not valid dynamically, imagine that you are told by person X, and accept, that *These scales are accurate* (the first premise). After John, a man of apparently normal weight, gets on the scales another person Y reads his weight and utters the second premise sentence, which you also accept. Will you necessarily come then to believe the conclusion that John weighs four hundred pounds? Hardly; for in accepting the second premise you would probably reject the first. This verifies by gedanken experiment the basic evidence statement that there exists an information state which, upon learning the premises in the order shown, does not come to believe the conclusion. On the other hand one sees that

These scales are accurate
If these scales are accurate then John weighs four hundred pounds
———————————————————— .
John weighs four hundred pounds

That is, as a *logical* implication the argument is valid. This is because no one could come to believe both premises *simultaneously* at a high credibility level without also believing the conclusion at a high credibility level.

It is readily verified for any logically independent a and b that under Hypothesis B, a succeeded by *If a then b* does not dynamically imply b, whereas they do logically imply b. In other words, Hypothesis B is once again consistent with the observed phenomena. Hypothesis A, of course, is not consistent with the observed failure of the dynamic implication, though it is consistent with the observation that the logical implication holds.

Similar comments apply to the classical rule of Modus Tollens. One observes in English the failure of the dynamic implication

That baby cannot walk yet.
If my eyes serve me correctly, it can walk.
* ———————————————————— ↓
My eyes do not serve me correctly.

This contrasts with the validity of the corresponding logical implication. Hypothesis B is again consistent with both observations, while Hypothesis A is consistent only with the second.

8.17 ORDER OF PREMISES

An example was given in the last section showing that the rule of Modus Ponens fails when interpreted as a dynamic implication from a followed by *If a then b* to b. But the dynamic implication seems more plausible when the premises are taken the other way around:

If these scales are accurate then John weighs four hundred pounds
These scales are accurate
———————————————————— ↓
John weighs four hundred pounds.

For if someone believed the first premise sufficiently strongly, and then in spite of the second premise's initial implausibility he heard and came to believe it, it seems as though he would indeed end up believing the conclusion. Thus the validity of this dynamic implication in English is dependent upon the order in which the premise sentences are received. Hypothesis B predicts this.

The significance of the order of the premises in a dynamic implication can also be observed in connection with the rule of Modus Tollens and the Hypothetical Syllogism, where again it is predicted by Hypothesis B. That the order of consistent premises could matter is of course completely inexplicable under Hypothesis A unless resort is made to *ad hoc* explanations.

8.18 Incompatible Conditionals

We turn now from implication to an examination of the other kinds of logical relationships, starting in this section with logical inconsistency or incompatibility.

In one of his last papers, Frank Plumpton Ramsey commented that if one person were to assert "If it rains, Cambridge will win," and another were to reply "If it rains, they will lose," the two would be regarded as disputants, and each would try to show grounds for his own belief and absence of grounds for his rival's (1931, p. 233). If one were to reformulate Ramsey's remark as a piece of logico-linguistic evidence, one could write

If it rains, Cambridge will win
If it rains, they will not win
———————————

to indicate the logical inconsistency of the two sentences. To judge whether these two sentences are indeed logically inconsistent in English, one must try to imagine a rational English speaker who is so informed that he believes both sentences at the same time. Unless one is able to conceive of someone in such a state, one has to regard this example at least tentatively as a genuine case of logical inconsistency.

So far as I know, it is true generally that for any pair *a* and *b* of logically independent simple sentences, *If a then b* and *If a then not-b* are logically inconsistent with English. Moreover the inconsistency is, for naive English speakers as opposed to logicians at least, rather striking and strongly felt. On the basis of Hypothesis A this fact is mystifying, for the two sentences are perfectly consistent according to the classical logic.

One might try to patch up Hypothesis A with some sort of *ad hoc* explanation of why the two sentences "seem" inconsistent even though they are not "really" so, and this is apparently what Ramsey was trying to do when he wrote (ibid., p. 247 n.)

If two people are arguing 'If p will q?' and are both in doubt as to p, they are adding p hypothetically to their stock of knowledge and arguing on that basis about q; so that in a sense, 'If p, q' and 'If $p, \bar{q}$' are contradictories.

But in the present theory there is a definite empirical criterion of inconsistency which this example satisfies, so the convenient way out of saying that the sentences merely "seem" inconsistent is not available. It is interesting, by the way, that the explanation which Ramsey seems to be invoking in the foregoing passage sounds very like a mentalistic formulation of our rule B1; i.e. the *ad hoc* explanation which Ramsey would add to the classical logic to account for the phenomenon in question consists in going over to Hypothesis B whenever one talks about that particular phenomenon. As an *ad hoc* explanation it is successful, for Hypothesis B does indeed predict that the sentences should be inconsistent. Of course, if Ramsey had adopted Hypothesis B in the first place, no *ad hoc* explanation would have been necessary.

8.19 SELF-CONTRADICTORY CONDITIONALS

The following passage is taken from an introductory logic textbook by D.A. Johnson (1963):

What would you think if your friend said, "If today is Monday, then today is not Monday?" Probably you would think he was silly since a given sentence is either true or it is false, but not both. The conclusion that a given sentence is both true *and* false is called a contradiction. . . .

What Johnson seems to be saying is that the English sentence in question is a contradiction – that it is logically self-contradictory, or inconsistent. In our notation, his claim would be written:

$$\frac{\textit{If today is Monday then today is not Monday}}{\quad}\,.$$

As a claim about English it seems to be correct, for it is impossible to imagine a rational English speaker having information that would cause him to attach any substantial credibility level to this sentence. So long as *Today is Monday* is accorded a probability greater than zero, and the senses of *today* and

Monday are held constant from antecedent to consequent, the sentence seems to the linguistic intuition as self-contradictory a statement as the English language has to offer.

What is ironic is that the quoted passage is from a textbook on the *classical* logic, a system of logic in which *If a then not-a* is not self-contradictory at all. To the contrary, in classical logic *If a then not-a* is logically consistent for all non-tautologous *a*. But apparently the intuitive feeling of self-contradiction surrounding *If a then not-a* is so strong in English speakers that they get confused and use it as an example of a logical contradiction even when they happen to be logicians attempting to illustrate the classical logic, in which it is not a contradiction.

This slip is not pointed out to embarrass Johnson, for I have observed the same confusion in conversations with other logicians as well. It is apparently an easy confusion to fall into, and that is the very point of the example. The fact that logicians tend in unprepared moments to call sentences of form *If a then not-a* self-contradictory even when talking about the classical logic is evidence of an especially striking kind of its self-contradictoriness in English, for in order to make this slip the English intuition of these logicians must be strong enough to overwhelm for a moment even their professional mathematical familiarity with the classical logic and its properties.

The logic of Hypothesis A being classical, sentences of form *If a then not-a* (where *a* is not logically valid) are logically consistent under it, contrary to English intuition. Under Hypothesis B, however, such sentences are always logically inconsistent.

8.20 ARISTOTLE'S SLIP

Those of us who as informants find *If a then not-a* to be self-contradictory in English can cite a venerable precedent in Aristotle, who apparently felt the same way about the Greek conditional. Aristotle wrote

> . . . it is impossible that *B* should necessarily be great if *A* is white, and that *B* should necessarily be great if *A* is not white. For if *B* is not great *A* cannot be white. But if, when *A* is not white, it is necessary that *B* should be great, it necessarily results that If *B* is not great, *B* itself is great. But this is impossible.

Lukasiewicz gives as a modern interpretation of this passage the following argument (1951, p. 50):

Although the example chosen by Aristotle is unfortunate, the sense of his argument is clear. In terms of modern logic it can be stated thus: Two implications of the form *If a then b* and *If not-a, then b* cannot be together true. For by the law of transposition [contraposition] we get from the first implication the premise *If not-b then not-a*, and this premise yields together with the second implication the conclusion *If not-b then b* by the law of the hypothetical syllogism. According to Aristotle this conclusion is impossible.

Lukasiewicz goes on to explain that Aristotle was mistaken in believing that *If not-b then b* is impossible. He also castigates another commentator for being so ignorant of logic as to agree with Aristotle on that point.

Putting questions of textual interpretation aside, let us assume for the sake of the discussion that Lukasiewicz' presentation of Aristotle's argument is a fair one, and examine that argument as Lukasiewicz reconstructed it. The argument could hardly be more suspicious, for it purports to prove that *If a then b* and *If not-a then b* are inconsistent, a result which is at odds with both Hypotheses A and B as well as with linguistic intuition. The challenge is, therefore, to pinpoint Aristotle's slip. According to Lukasiewicz the flaw in the argument comes in Aristotle's assumption that *If not-b then b* is self-contradictory. This is indeed the flaw in the argument provided the argument is in a language described by Hypothesis A. But if one is reasoning under Hypothesis B, *If not-b then b* is self-contradictory after all, just as Aristotle assumed, and one must look for the flaw elsewhere.

The flaw is not hard to find. Aristotle's argument assumes both the law of contraposition and the law of the hypothetical syllogism, neither of which is valid under Hypothesis B. And upon examining the argument in detail, one finds that the law of the hypothetical syllogism has been applied under exactly those conditions in which Hypothesis B predicts that it will lead to absurdities. Thus Hypothesis B resolves the difficulty without the necessity of renouncing Aristotle's intuitively plausible assumption that *If not-b then b* is impossible.

In fairness to Aristotle it should be added that a thorough logico-linguistic analysis of classical Greek might conceivably show that language to follow neither Hypothesis A nor Hypothesis B, but instead some third set of rules under which Aristotle's whole argument turns out to be valid after all. If so, we would have an instance of an argument which is either valid, or invalid with Flaw One, or invalid with Flaw Two, depending upon whether it is read in the original Greek, in a language governed by Hypothesis A, or in a language governed by Hypothesis B.

8.21 INCOMPLETENESS OF THE RULES GOVERNING CONDITIONALS

In English, conditionals with self-contradictory antecedents sound very odd; the hearer would have a hard time knowing what to make of them. Consider for instance

If it is raining and it is not raining, then the Republicans will win.

Whatever this peculiar sentence might mean, unindoctrinated native speakers of English would certainly not be unanimous in saying that it would be believed with certainty by all rational speakers no matter what their factual information.

This datum conflicts with Hypothesis A, under which a sentence of form *If* [*a and not-a*] *then b* ought to be a tautology like any other tautology. One might defend Hypothesis A by bringing in the *ad hoc* explanation that an indicative conditional is normally never used with a contradictory antecedent, and this explanation has a certain amount of plausibility to it. But *why* is an indicative conditional never used with a logical contradiction as an antecedent? Hypothesis A by itself gives us no clue, though here again one might bring in *ad hoc* explanations.

Under Hypothesis B it is obvious why an indicative conditional with a contradiction as its antecedent seems odd and hard to interpret. It seems odd because the credibility of such a sentence is simply undefined: Rule B1 defines it only for cases in which the antecedent can be learned. According to Hypothesis B the rules of English fail to cover this situation, in consequence of which there is little wonder that the sentence seems inscrutable. The structuralizations of English are simply incomplete in this respect.

8.22 LOGICALLY DISJUNCT CONDITIONALS

A type of logical relationship for which the gathering of evidence has not yet been illustrated is the oft-neglected relationship of logical disjunctness. By our definitions sentences are logically disjunct if it would be impossible to be so informed as to assign a low credibility to all of them simultaneously. As an example of a pair of sentences that are *not* logically disjunct in English we have

* ---

There are human beings living on the surface of the sun.
If there are human beings living on the surface of the sun, then they suffer mainly from the extreme cold.

People so informed as to strongly disbelieve both sentences are all around us, so there is nothing difficult about imagining their existence.

Hypothesis A predicts that the sentences in question must be logically disjunct, and suggest no explanation for why they do not seem so. But Hypothesis B successfully predicts that they are not logically disjunct.

8.23 NEGATIONS OF CONDITIONALS

Up to this point we have examined the English conditional only in its occurrences as the main connective of a sentence. We turn next to sentences which are themselves built up out of conditional components, starting in this section with negations of conditionals, and continuing in the remaining sections to examine conjunctions of conditionals, conditionals whose components themselves involve conditionals, and disjunctions of conditionals.

It is sometimes asserted that conditionals are never negated in colloquial usage, but that does not seem to be so. Negated conditionals are unusual, but can occur in contexts such as the following imaginary conversation among three businessmen:

X: *If John retires before he is sixty, he will still retain his pension benefits*

Y: *That's not so.*

Z: *Pardon me, what's not so?*

Y: *It's not so that if Jones retires before sixty, he will retain his pension benefits.*

And on those rare occasions when they can occur, negated conditionals seem to carry meaning, so that it is not out of the question to try to investigate their properties logico-linguistically.

Here are some examples of English arguments involving negations of conditionals which would never pass the informant test for logical implication:

It is not the case that if the peace treaty is signed, war will be avoided.
* ———
The peace treaty will be signed.

It isn't true that if he breaks a mirror he will have bad luck.
* ———
If he doesn't break a mirror he will have bad luck.

It is not the case that if we follow that road we will reach the city.
* ———
We will not reach the city.

It is not true that if she is over forty she is still young.

* ___

If she is still young she is over forty.

These absurd arguments are predicted to be valid under the classical logic of Hypothesis A, but invalid under Hypothesis B. This situation remains unaltered if the arguments are interpreted as dynamic instead of logical implications.

The reverse situation can also occur. Here is an argument that seems plausible enough in colloquial English:

If production is rising, profits are not declining.

___

It is not the case that if production is rising profits are declining.

This acceptable instance of logical implication is invalid according to Hypothesis A, valid according to Hypothesis B. Reinterpreting it dynamically does not change anything.

There are also some mixed situations. Tested in terms of dynamic implication, the following arguments do not fare too badly:

Today is not Wednesday.

___ ↓

It is not the case that if today is the thirteenth, today is Wednesday.

Either the smallest integer in the set is non-prime, or the greatest integer in it is non-prime.

___ ↓

It is not the case that if the set contains just two integers, the smallest integer in it is prime and the greatest integer in it is prime.

On the other hand, they do not fare as well when considered as logical implications. To think of a counterinstance to the first argument, for example, one need only imagine a person who is fairly sure it is still Tuesday the twelfth, though near midnight. Hypothesis A predicts that the arguments are invalid both logically and dynamically. Hypothesis B predicts that they are invalid logically but valid dynamically, as our intuition as informants seems to suggest.

According to Hypothesis A, *Not-*[*if a then b*] logically implies *a and not-b*, resulting in absurd English arguments. Under Hypothesis B it merely implies *if a then not-b*, which is much weaker. But even the latter conclusion may be too strong sometimes. For while one might be inclined to accept

It is not the case that if Jones retires before sixty he will retain his pension benefits.

If Jones retires before sixty he will not retain his pension benefits.

it seems reasonable to challenge

It is not the case that if Jones' car is gone he is out.

* ---

If Jones' car is gone he is not out.

It is unclear (to me) just what is going on in examples like these. Perhaps *It is not the case that* when followed by a conditional statement is sometimes understood to mean *It is not necessarily the case that.* Or perhaps postulate B4 is indeed too strong. Or perhaps negations of whole conditional statements, being rare in English, have an interpretation which is idiosyncratic and simply unsettled at this point in the evolution of the language.

8.24 CONJUNCTIONS OF CONDITIONALS

The following conjunction of conditionals seems self-contradictory in English:

If John loses he will try again, and, if he loses he will not try again.

The usual gedanken experimentation reveals that English has as a basic evidence property that no rational speaker could be so informed as to believe the sentence in question: if someone were to assert such a sentence, one would look for signs of irrationality on his part, for a special hypostatic usage, for shifts in the intended denotation of *he*, or something similar. Now according to Hypothesis A, a sentence of form *[If a then b] and [if a then not-b]* is not logically self-contradictory at all. But under Hypothesis B a sentence of this form is indeed predicted to be logically self-contradictory, in consonance with the results of informant technique.

As a slightly more complicated bit of evidence, most English speakers would probably agree that there is something logically inconsistent about the following pair of sentences:

It will not both rain and shine.
If the barometer drops it will rain, and, if the barometer drops it will shine.

This inconsistency too is predicted by Hypothesis B but not by Hypothesis A.

It is a peculiar feature of Hypothesis B that under it conjunction is not always commutative. For example, the credibility level at which a sentence of form *a and [if not-a then b]* is believed is just the product of the credibilities of its conjuncts, whereas the credibility level of *[If not-a then b] and a* is less straightforward and cannot be determined from B1-B5 alone. Is there any trace of a corresponding noncommutativity in English? The speech phenomena involved are rather delicate, sentence types such as these being used only rarely in everyday discourse. Nevertheless there may be occasions when assertions of the first form are made, as in

> *That bird is a nuthatch, and if it isn't a nuthatch it's a species that looks a lot like a nuthatch.*

What the speaker seems to be saying is that the bird is very likely a nuthatch, but in the unlikely event that it isn't it is probably of a species that looks similar. This interpretation is in consonance with the way the credibility of the sentence is computed under Hypothesis B.

Now consider the commuted form of the sentence:

> *If that bird isn't a nuthatch it's a species that looks a lot like a nuthatch, and it's a nuthatch.*

This sounds rather odd stylistically and its meaning is slightly perplexing. It is hard to think of circumstances under which it might naturally be uttered, and its intent is not as clear as the former version of the statement. Thus the fact that under Hypothesis B the credibility conditions of the second sentence are different from those of the first, and in particular more complicated and obscure, does not run counter to the linguistic intuition and may even be reflected to some extent in the way the sentences are used (or in the case of the second sentence, avoided) in ordinary discourse.

Hypothesis A, of course, comes nowhere near an appropriate handling of either sentence. According to Hypothesis A, both *a and [if not-a then b]* and *[if not-a then b] and a* are logically equivalent to *a* alone. Thus Hypothesis A supports the implication

> *That bird is a nuthatch.*
>
> * ————————————————————————
>
> *That bird is a nuthatch, and if it isn't a nuthatch it's a species that looks a lot like a nuthatch.*

and also the converse argument, whereas linguistic intuition leads one to believe that one sentence says more than the other.

Another instance of noncommutativity involves the sentence forms [*Not-b*] *and* [*if a then b*] versus [*If a then b*] *and* [*not-b*]. Under Hypothesis B, the first of these is logically self-contradictory, while the second has an obscure credibility level which cannot be determined from B1-B5 alone. Instances of these sentence forms in English would be

He will not come and if you ask him he will come

and

If you ask him he will come, and, he will not come.

Both these sentences sound odd and unlikely, and Hypothesis B is successful to the extent that it predicts this oddness, on the grounds of self-contradictoriness for the first sentence and incompleteness of the rules of English for the second. On the other hand the success of Hypothesis B in so predicting has to be qualified insofar as it explains the oddity of the two sentences on different grounds, while the linguistic intuition finds little to distinguish between the two. The best summary of the situation is, perhaps, that these two sentences do indeed provide a tentative bit of counterevidence against Hypothesis B, but that it is by no means clear and striking counterevidence because it concerns sentence forms which do not ordinarily occur in colloquial discourse.

Hypothesis A is once again wide of the mark. According to Hypothesis A both sentences should be equivalent to

He will not come and you will not ask him.

8.25 CONDITIONALS CONTAINING OTHER CONDITIONALS

Imagine for a moment that you happen to know that Tom, Dick, and Harry are all such large men that none of them has much room to spare when standing in a phone booth. Then you might well find the first of these sentences credible but not the second:

It is not the case both that Tom is in that phone booth and that Dick is in it too.

* ---

If Tom is in that phone booth, then if Dick is in it too so is Harry.

The little gedanken experiment you just performed shows the argument not to be a valid logical implication in English. This experimental evidence stands in conflict with Hypothesis A, under whose classical logic the inference is perfectly valid. But it tends to further corroborate Hypothesis B according to whose rules the argument is invalid. The invalidity under B is due essentially to the fact that no matter how unlikely it is thought to be that both Tom and Dick are in the phone booth, it could always be deemed less likely still that all three men are in it.

Next consider the sentence

* ___

If they can't afford it, then if they can afford it they should buy it immediately.

Would this peculiar assertion necessarily be believed by every rational English speaker no matter what his information? It would seem not, as Hypothesis B predicts, though according to Hypothesis A the sentence ought to be a tautology.

Here is a sentence of slightly different logical form which sounds as though it could never be rationally asserted by anyone at all:

If Smith is an American citizen, then if he used to be a boy scout he is not an American citizen.

___

This unbelievable sentence is logically self-contradictory under Hypothesis B but not under A. Under Hypothesis A it is consistent and is moreover supposed to imply

If Smith was never a boy scout then he is not an American citizen

which is absurd.

This gives us some idea of the contrast between the two hypotheses with respect to conditional statements whose consequents are themselves conditionals. Turning to conditional statements whose antecedents are themselves conditionals, one finds the evidence scantier, for in colloquial usage conditionals seem to be nested in this direction less often than in the other. Under Hypothesis B this tendency is explainable in part by the fact that the credibility of a sentence of form *If a then* [*if b then c*] is wholly determined by rules B1-B5, whereas the credibility of *If* [*if a then b*] *then c* involves some indeterminacy. But whether this explanation of their rarity is correct or not, left-nested conditionals do seem to have a certain amount of meaning in at least some instances, and so are worth examining briefly.

For a start let us consider the inference

Grandmother is outdoors and she is not a bit cold.

* ___

If grandmother is a bit cold if she is outdoors, then she will move into the shade.

The second sentence is somewhat awkward stylistically, but to the extent that it can be understood it seems to say something other than what is implied by the first sentence; a rational speaker could believe the first sentence and not the second. We therefore tentatively classify the argument as invalid for English. Classically it is valid, but under Hypothesis B it is not; hence Hypothesis B is again found to be more satisfactory.

A sentence which seems hard to believe under any factual assumption is:

If that lion is unfriendly if provoked, then he is friendly and he is provoked.

The sentence is not logically self-contradictory under Hypothesis A, but it is under B.

A sentence that simply sounds muddled is

If she didn't do it if she did, then she didn't.

Sentences of this form are logically valid under Hypothesis A – everyone is always supposed to believe them as truths of logic. But under Hypothesis B they are not logically valid; in fact there is no state in which the credibility function is even defined for them. The muddled character of such sentences is thus explainable under Hypothesis B on the grounds that the rules of English simply do not cover that particular sentence form.

Here is a somewhat more involved example:

If it is the case that if Jones passes logic he will graduate, then he will graduate.

* ---

If Jones does not pass logic he will graduate.

The argument, though classically valid, is highly suspicious. The purported inference fails under Hypothesis B with the premise having a credibility determinable only on extralinguistic grounds.

Another classical whopper which is blocked by Hypothesis B in a similar manner is:

If I'm apt to get hurt if I try it, I won't try it.

* ---

If I try it then I'm not apt to get hurt.

Now let us inspect an argument involving a conditional whose consequent is a negated conditional. It is about Goldbach's famous unproved conjecture that every even number is the sum of two prime numbers:

If Goldbach's conjecture is correct, then it is false that if the mayor's telephone number is even it cannot be represented as the sum of two primes.

* ───────────────────────────────

If the mayor's telephone number is not an even number, Goldbach's conjecture is not correct.

This is an especially droll bit of classical reasoning, for it leads from a believable premise to the startling discovery that Goldbach's conjectures can be disproved simply by showing the mayor's telephone number to be odd! It is valid under Hypothesis A, invalid under B.

To wind up the survey of nested conditionals we consider a couple of arguments which are invalid classically but which nevertheless have a certain plausibility in English:

If Titan is without an atmosphere, it has no life.

─────────────────────────────── ↓

If Titan is without an atmosphere, it is not true that if it is temperate it has life.

If we rent the car, we will not need our own insurance.

─────────────────────────────── ↓

It is not true that if we rent the car, we will need our own insurance if we want to be protected financially.

Invalid both logically and dynamically under Hypothesis A, these arguments have an in-between status under Hypothesis B – they are invalid logically but valid dynamically. Their analysis under Hypothesis B is similar to that of the second paradox of material implication (Section 8.15).

8.26 LEWIS CARROLL'S BARBERSHOP PARADOX

At a time when modern symbolic logic was still in its infancy, the author of *Alice's Adventures in Wonderland* published a 'paradox' which has been the subject of much controversy since (Carroll 1894). The paradox concerns three barbers – Allen, Brown, and Carr by name – who run their barbershop according to the rules that: (i) at all times at least one of the three must be in the shop; and (ii) whenever Allen goes out, Brown must accompany him. Lewis Carroll noticed that on the basis of these rules the following statements can be made:

(i) *If Carr is out, then if Allen is out Brown is not out.*
(ii) *If Allen is out, Brown is out.*

Now Lewis Carroll believed that any two statements of the form *If a then b* and *If a then not-b* must be logically incompatible with each other. But this, he observed, makes (ii) contradict the consequent of (i), yielding (by a form of Modus Tollens) the paradoxical conclusion that Carr must be in.

Let us analyze this paradox on logico-linguistic principles. That the conclusion that Carr must be in seems paradoxical is a logico-linguistic fact of English which may be recorded thus:

> *If Carr is out, then if Allen is out Brown is not out.*
> *If Allen is out, Brown is out.*
> * ———————————————
> *Carr is not out.*

Lewis Carroll's description of the rules of the barbershop amounts to a specification of an information state in which the premises of this argument have high credibility but not the conclusion. Thus Carroll performed and led his readers to perform the gedanken experiment showing the argument to be invalid in English. So far, no problem. The problem arises only with Carroll's enunciation of certain alleged logical principles according to which, so he claimed, the argument *ought* to be valid. Which of these principles is faulty?

Under Hypothesis A (and this is the standard classical resolution of the paradox), the fault in Carroll's logic lies in the mistaken assumption that sentences of form *If a then b* and *If a then not-b* are logically incompatible. Since these are not really incompatible (classically), Carroll's chain of reasoning is broken and there is no reason to suppose that the argument should be valid. Thus Hypothesis A succeeds in avoiding the difficulty, and provides a resolution of the paradox – of sorts.

But is the resolution a satisfactory one? If Lewis Carroll had stated his paradox in the language of truth-functional logic, this resolution might be accounted satisfactory. Unfortunately for the classicists, though, he didn't; he stated it in English. English is a language in which *If a then b* and *If a then not-b* are logically incompatible after all as we saw earlier (Section 8.18), i.e. we have

> *If Allen is out Brown is out.*
> *If Allen is out Brown is not out.*
> ———————————————

as a legitimate instance of logical inconsistency in English. And if this pair of sentences is indeed logically inconsistent, as Lewis Carroll claimed and as our linguistic intuition confirms, the classical escape route from the paradox is cut off. The escape is itself faulty in claiming to find a fault where there is no fault.

Hypothesis B offers a more satisfactory resolution of the paradox. Like Hypothesis A, Hypothesis B rejects Carroll's paradoxical reasoning, but it rejects it for different and (in English) better reasons. Under Hypothesis B Carroll was quite correct in assuming that *If a then b* is incompatible with *If a then not-b*; the mistake in his reasoning lay not there but rather in assuming that the rule of Modus Tollens holds good in English in the form in which he applied it. It was noted earlier (Section 8.16) that Modus Tollens can fail dynamically even for conditionals with simple consequents. The Barbershop Paradox illustrates the fact that, as Hypothesis B predicts, it can fail logically too when the consequent is itself a conditional.

8.27 Disjuctions of Conditionals

When one starts to examine disjunctions of conditionals one finds an abundance of argument forms which ought to be valid according to Hypothesis A, but which are weird in English. Here for instance are a couple of sentences which are supposed to be logically valid according to Hypothesis A but which don't sound tautologous at all.

* ______________________________

If she's rich she's happy, or else if she's happy she's rich.

* ______________________________

Either if this figure is rectangular and not equal sided it is a square, or if this figure is not rectangular and is equal sided it is a square.

They are not logically valid under Hypothesis B.

Some curious one-premise and two-premise arguments obtained under Hypothesis A follow.

If John is in Paris, then he is in France.
If he is in Istanbul, then he is in Turkey.
* ______________________________
If John is in Paris he is in Turkey, or, if he is in Istanbul he is in France.

If the temperature drops, it will snow.
* ______________________________
It will snow, or if the temperature drops it will rain.

If the main switch is on and the auxiliary switch is on, then the motor is on.

*——————————————————————

If the main switch is on the motor is on, or, if the auxiliary switch is on the motor is on.

If Jones comes from Mississippi, he is a southerner.

*——————————————————————

If Jones rides a bicycle to work he is a southerner, or, if he comes from Mississippi he owns a station wagon.

If it rains, then it will not snow.

*——————————————————————

If it rains the game will continue, or if it snows the game will continue.

If Albert's age is greater than twenty and less than twenty-three, then Albert is either twenty-one or twenty-two.

*——————————————————————

If Albert's age is greater than twenty then Albert is twenty-two, or if his age is less than twenty-three then he is twenty-one.

Under Hypothesis B none of these arguments is valid. Evidently Hypothesis B successfully blocks many unwanted inferences involving disjoined conditionals that Hypothesis A would allow.

As was found to be the case with conjunctions of conditionals (Section 8.24), so with disjunctions of conditionals, Hypothesis B sometimes predicts noncommutativity. For example, under Hypothesis B *a or* [*if not-a then b*] has a readily determined credibility value which turns out to be the same as that of the simple disjunction *a or b*, while the credibility of [*If not-a then b*] *or a* is left undetermined by B1-B5. This situation seems to be reflected to some extent in colloquial usage. A sentence like

I will be there, or if I'm not there my partner will be there.

sounds fairly natural and seems equivalent to the simple disjunction

I will be there or my partner will be there.

as predicted by Hypothesis B. On the other hand, and also as predicted by B,

If I'm not there my partner will be there, or I will be there

sounds strained, and seems to mean if anything something rather different. Hypothesis A, of course, predicts perfect commutativity for '*or*' under all circumstances.

A striking phenomenon to be observed in connection with the English disjunction is that when it is used to disjoin two conditional statements it sometimes has the force of a conjunction, i.e. *or* sometimes seems to mean *and*. It seems to be used in this fashion most frequently in very casual or informal circumstances, and in conversational speech more than in writing. It may even be dialectal, making any linguistic evidence gathered about it rather tentative and delicate, but the phenomenon is so surprising that it is worth examining in spite of its elusive character. Here are three sample arguments in which *or* seems to act like a conjunction:

> *If that player trades pawns he will lose, or if he doesn't trade pawns he will still lose.*
>
> ---
>
> *That player will lose.*

> *If the new manager is successful, he will be promoted, or if he is not successful, he will be transferred.*
>
> ---
>
> *Either the new manager will be promoted or he will be transferred.*

> *If the water supply is off then she cannot cook dinner, or if the gas supply is off she cannot cook dinner.*
>
> ---
>
> *If either the water supply is off or the gas supply is off, she cannot cook dinner.*

These arguments may not carry complete conviction for every English speaker, but the interpretation of the premises under which they are valid seems natural enough under at least one manner of speaking.

Neither Hypothesis A nor Hypothesis B successfully predicts the validity of these arguments. Hypothesis A is way off, putting the premises of the first two examples in the category of tautologies, which is absurd; and Hypothesis B, though it does not make that particular blunder, nevertheless fails to assign the premises the conjunctive senses they seem to have intuitively. How then is the conjunctive interpretation of the *or*'s to be explained?

Under Hypothesis A there does not seem to be an easy explanation, but under B the phenomenon can be accounted for in an interesting way.

Consider the disjunctive sentence in the first of the trio of arguments, which is of form [*If a then b*] *or* [*if not-a then b*]. Upon studying the way the rules B1 and B3 apply to sentences of this form, a remarkable fact emerges: under Hypothesis B, it is impossible for the disjunctive sentence in question to have high credibility unless at least one of its disjunctive parts also has high credibility. (This contrasts with disjunctions of simple sentences of form *a or b*, in which the sentence as a whole can have high credibility even though neither *a* nor *b* alone has a very high credibility.) Thus this particular disjunction cannot honestly be asserted except under one of the following pair of circumstances: (a) one of the disjuncts is believed but not the other; or (b) both disjuncts are believed. But the hearer can reasonably rule out possibility (a) on the higher pragmatic grounds that if the speaker believed one of the disjuncts but not the other, he would have simply uttered the one he believed and omitted all mention of the other. Hence the hearer can deduce that (b) must be the case, which of course gives the *or* the thrust of an *and*. The other two examples can be reconciled with Hypothesis B along similar lines.

As a final bit of evidence we present another intuitively plausible argument involving a casual and possibly dialectal use of *or*:

> *The ceremony will be held in the main hall, or if the weather warms up, it will be held on the green.*
>
> ---
>
> *Either the ceremony will be held in the main hall or it will be held on the green.*

The implication is invalid under Hypothesis B as well as A. This time, moreover, it is not so easy to reconcile the observation with Hypothesis B. The argument seems to be a genuine piece of counterevidence against Hypothesis B. It and its substitution instances are in fact the most clear-cut bits of evidence against Hypothesis B of which I am currently aware.

8.28 CONCLUSIONS ABOUT *IF–THEN*

This completes the illustrative analysis. The stopping point is admittedly arbitrary, for it would have been possible to extend the study to include sentence forms with a still greater depth of nesting of the connectives; to bring in basic evidence properties other than the logical and quasi-logical relationships; to check the evidence out on more informants; to introduce more orderly and unbiased ways of selecting the argument forms about which evidence is gathered; and so on. But even as it stands the study would seem to support the following conclusions.

1. Hypothesis A can safely be rejected. The counterevidence to it has now piled up to the point where it would be absurd to try to save it by

introducing *ad hoc* explanations of individual examples. (Variants of Hypothesis A however, or Hypothesis A butressed by some special rules of interpretation or application, are not necessarily entirely ruled out.)

It might be objected that no one ever seriously advocated the material conditional as a formalization of *if–then* in exactly the form in which it was tested here as Hypothesis A. Perhaps not, but those who defend an association between *if–then* and the material conditional should be obliged to state precisely what hypothesis they *do* advocate, to spell it out in enough detail so that it can be tested out against the evidence, and then test it.

2. Although they have not been examined specifically here, grave doubt is cast on many of the other claims that are sometimes made about *if–then*, for example that it is formalizable by the intuitionistic conditional, by strict implication, or by the conditional of a standard *n*-valued logic. Indeed, many of these systems are fragments of the classical logic in the sense that arguments valid in them are valid classically, and this alone brings much of the evidence against the material conditional to bear against them too.

3. The evidence examined so far, though not conclusive, supports Hypothesis B fairly strongly. Of the English arguments surveyed, few were unearthed that seemed to weigh at all heavily against Hypothesis B. The vast majority supported B, and this in spite of the fact that most were 'difficult' pieces of evidence in the sense that Hypothesis A was unable to explain them. Of course, as with any scientific hypothesis it is always possible that more serious counterevidence to Hypothesis B will be discovered in the future.

4. Just as the evidence reviewed makes it more plausible that *if–then* is governed by the rules of Hypothesis B (or rules something like them), so the same evidence makes it more plausible that the general model needed for all natural language analysis obeys the working assumptions underlying Hypothesis B (or something like them). Among the working assumptions were semantic structuralizability in the extended sense and the probability-weighting of the semantically structuralizable sublanguage serving as the basis for the extension. The fact that probability-weighting, for example, is an intrinsic part of a hypothesis that has been tested and found reasonably successful supports the conclusion that at least some natural language rules are essentially probabilistic in character. The implications for the nature of human thought are intriguing to say the least.

8.29 FURTHER CASE STUDIES

In addition to *and*, *or*, *not*, and *if–then* I have examined several other different sorts of English structures using a similar logico-linguistic methodology. In particular I have analyzed elsewhere in some detail the indefinite article of English (Cooper 1976). Without attempting to summarize the results of that study here, it may be said that they support the use of an underlying apparatus similar to the one upon which Hypothesis B was based. In other words, as in the *if–then* study the most descriptively accurate hypothesis turned out to be one which assumed semantic structuralizability in the extended sense, using probability weightings and pragmatic rules formulated in terms of a belief function rather than a classical semantical satisfaction relation. These results tend to reinforce Conclusion 4 of the foregoing study in suggesting that pragmatic rules of the kind whose expression is made possible by the general theory presented here are needed to describe at least some important and ubiquitous natural-language devices.

8.30 CONCLUDING REMARK

It has been an underlying theme throughout that the exploration of the logic of natural language is not just a philosophical but also a *scientific* enterprise. It involves evidence-gathering, hypothesis-testing, and all the other trappings essential to the scientific approach. I hope this chapter has suggested, if nothing else, that the scientific approach is appropriate to the discovery of natural-language logic – that 'descriptive logic' is not a contradiction in terms. The gap between theory and observation is not unbridgeable in this field. Reasonably precise hypotheses *can* be framed, observational evidence *can* be gathered to test them, and generally speaking one is encouraged in the hope that logico-linguistics may be a viable science.

CHAPTER 9

PROBLEM AREAS AND COMPUTER APPLICATIONS

It will be said that the idea of a unified science of language and logic raises more problems than it solves. While this is no doubt true, it is not necessarily bad. The important question is whether they are legitimate questions to raise – whether in fact they *should* be raised. In this chapter some miscellaneous problem areas will be touched upon and some potential computer applications indicated.

9.1 CHOICE OF LINGUISTIC UNIT

When the logician-linguist sets out to analyze a language, the first practical decision he must make is the choice of a suitable linguistic unit for the description. Logico-linguistic theory tells him only that sentences are to be represented as finite sequences of linguistic entities of some sort; it does not specify *what* sort. Should they be taken to be sequences of phones, phonemes, morphemes, sememes, words, or what?

Obviously the sequences of entities in the formal sentence set, whatever they are, must stand in one–one correspondence with the sentences of the observed language. Beyond this the only absolute requirement is that this correspondence be recognizable by the analyst in order that observational consequences of his language description can be compared readily against observational data. To make the correspondence obvious the natural thing to do is to let the structure of the mathematical sequences match the temporal or spatial structure of the observational sentences. Basically, any choice of linguistic unit which achieves this is a candidate for the logico-linguistic description of the language.

The choice of linguistic unit will of course affect the character of the language description. If a low-level unit is chosen – say phones or phonemes for a spoken language or characters or graphemes for a written – the correspondence between an observed sentence and the abstract mathematical sequence representing it will be fairly immediate, involving phonology or graphology only; there will be little risk that the analyst will associate the wrong sequence with an observed sentence when testing out his description, and later users of the description should find the relationship relatively straightforward and mechanical. If on the other hand a higher-level unit such as the morpheme or word is chosen, the connection between formal description and observable reality also involves morphology and so is a chain with more theoretical links; the reader or verifier of the description is called upon to fill in larger gaps when applying the description. Hence a low-level unit is preferable from the point of view of the completeness of the description and minimizing the chance of misapplication, while a well-chosen higher-level unit might make it simpler, shorter, and perhaps in some ways more natural. A scientifically complete language description would be as low-level as possible, using as units basic speech sounds or even raw acoustic or articulatory patterns, but compromises with this ideal may be made for the sake of a briefer description or perhaps simply because the analyst does not happen to be interested in the very low-level structure of the language.

It seems plausible that an astute choice of higher-level representation might not only shorten the syntactic part of a logico-linguistic language description but also simplify greatly the later parts of the description dealing with the state set, learning operation, and credibility function. Even if a low-level unit is decided upon, the initial stages of the description are apt for this reason to consist of rules for transforming lower-level sentence representations into higher. If for example a description takes phonemes as its fundamental or observational unit but uses morphemic representations in its post-syntactic parts, a set of morpho-phonemic rules must be included somewhere in the description. This kind of description seems the most desirable; the low-level observational terms (phones, etc.) make it a full theory of the language with clear empirical content while the higher-level theoretical terms (morphemes, sememes, etc.) lend it the explanatory power and simplicity one expects from a good scientific hypothesis.

A language description with low-level observables can therefore be expected to include what have traditionally been called the 'structural' rules of the language, from the level at which sentences are represented for observational purposes on up. These rules must somehow specify, for every sentence-representing sequence on the observational level, a unique higher-level representation. Furthermore two lower-level sequences must be allowed to map into the same higher-level sequence only in case the two are absolutely synonymous (§ 6.4). Presumably the rules that would accomplish this are much like those already familiar to structural linguists, and it is in this sense that logico-linguistics includes all of traditional structural linguistics.

If structural linguistics is needed for logico-linguistics it can also be argued that the need is mutual. One of the most controversial questions of structural linguistics has been: What higher-level units should be introduced and what criteria must they satisfy? Some new light is shed on the issue by the logico-linguistic perspective within which the only ultimate criterion is seen to be descriptive convenience: higher-level units may be defined in any way which renders the logico-linguistic description as a whole simple and natural, so long as descriptive accuracy is preserved. A related question is: How should one choose among the various competing schools of structural language analysis? The logico-linguistic answer is that if two structural methods account for the observational data equally well on the syntactic level, the only important consideration which remains is the issue of which facilitates the remainder of the logico-linguistic description. If both are equally viable as vehicles for the post-syntactic stages of the description, there may be no grounds left for choosing between them except those of taste.

There are also some more specific issues of structural linguistics to which logico-linguistics may offer further clarification. Examples are the traditional structural notions of 'functional equivalence' and 'free variation' (see e.g. Lyons 1969) which can be given a precise definition only with the help of a suitable criterion of synonomy such as logico-linguistics offers.

9.2 Ambiguity

It will be recalled that the Input Selector has the task of interpreting the sentences it submits to the Language Automaton in addition to assigning credibility weights to them. For languages containing ambiguous sentences we may imagine this task to be carried out by a special component of the Input Selector called the 'Disambiguator'. How the Disambiguator works is of indirect concern only here since disambiguation is part of higher pragmatics. Presumably it starts to operate at an early stage of the input selection process, for a credibility level can hardly be assigned to a sentence until its probable intended meaning has been decided upon. Presumably too the Disambiguator frequently consults the Language Automaton by submitting possible interpretations of ambiguous sentences to it as test inputs, for level of prior personal belief in a statement is one of the factors to be considered when comparing it with other possible interpretations of an ambiguous sentence. But the hows and whys of disambiguation do not matter to the logico-linguistic analysis of a language so long as the disambiguation takes place, and no one doubts that it does.

What *is* necessary for logico-lingusitic analysis is to settle upon suitable conventions for the form of the disambiguated sentences received by the

Language Automaton. The conventions can be chosen arbitrarily since the interpreted sentences are only indirect observables. Any reasonable way of modifying or extending an ordinary sentence-representing sequence will do so long as it yields a distinct new sequence for each possible interpretation of the sentence. Of course, if the interpreted sentences can in addition be made to differ in structure from the uninterpreted in ways indicative of the associated interpretation, this would be of at least mnemonic value.

Ambiguities are roughly classifiable as either syntactic, semantic, or due to context-dependence or deixis. If the sentence set of a language under analysis is described by conventional phrase-structure or transformational rules, the way in which a syntactic ambiguity has been broken by the Disambiguator can be indicated conveniently by including in the interpreted form of the sentence all or part of the sentence's derivational history. To use a stock example, the syntactic ambiguity in the sentence

Benjamin Franklin appreciated old men and women

can be removed by rewriting it as either

Benjamin Franklin appreciated [*old* [*men and women*]]

or

Benjamin Franklin appreciated [[*old men*] *and* [*women*]].

The brackets, which in traditional terms are partial indicators of a derivational tree, have significance in the logico-linguistic context as indicators of how the Language Automaton is to interpret the sentence. The bracketing convention was of course the device used to break structural ambiguities in the schemata examined in the case study of the last chapter.

As another example of syntactic ambiguity,

It is swimming now

could have as its interpreted form either

It is swimming/V *now*

or

It is swimming/N *now*

depending on whether the sentence is interpreted as answering such questions as "What is that marine creature doing?" or "What is your favourite athletic activity these days?" Such examples are of course oversimplified for the sake of the illustration, and deal with only one ambiguity per sentence where there may actually be multiple ambiguities The point is simply that conventional syntactic devices such as bracketing and form-class labelling, in addition to their customary place in the syntactic stages of a language description, can in a logico-linguistic analysis be made to serve another, interpretive, function in the case of syntactically ambiguous sentences. The ambiguity-breaking power of traditional syntactic analysis enters the logico-linguistic picture as a possible apparatus for constructing the interpreted sentences postulated as the product of the Disambiguator, in the case of syntactically ambiguous sentences at least.

Logically there is no necessity to include *all* of the derivational history of a sentence in the interpreted form of it that reaches the Language Automaton, but only enough to break its ambiguities. However, more interpretive data than is needed does no particular harm and merely adds some redundancy. The redundancy may even have its advantages in simplifying later parts of the logico-linguistic description, so there is a case for including much or even all of the derivational history ('deep structure', 'phrase-marker form', 'dependency-structure', etc., depending on the school of syntactic analysis) in the interpreted sentence forms. Remembering too the potential advantages of using higher-level linguistic units as theoretical expressions in the description (§ 9.1), we can now see at least three ways in which the linguistic research of the last few decades bears on logico-linguistics: first, it suggests ways of translating low-level sentence representations into higher-level ones in ways which simplify not only the syntactic rules of formation but also the later logico-linguistic rules; second, it allows the expansion of sentences into forms displaying their syntactic structure more explicitly, resulting in possible further simplifications of the logico-linguistic rules; and finally it suggests syntactic-ambiguity-free forms for interpreted sentences. Far from competing with current programs of linguistic research, logico-linguistic objectives would seem to complement them and lend sharper focus to those aspects of them that extend beyond syntax.

Another kind of ambiguity, commonly called 'semantic ambiguity', has to do with homonymy or the multiple meaning problem at the word or morpheme level. Consider for example the sentence

Richard Nixon put Henry Kissinger in his cabinet.

English dictionaries tend to list the 'cupboard' sense of *cabinet* as Meaning *a* and the 'advisory council' sense as Meaning *b*. This suggests recognizing on the morphemic level two distinct morphemes {cabinet^a} and {cabinet^b}. With all other data about derivational history omitted, the interpreted form of the sentence would then be something like

$$\{Richard\} + \{Nixon\} + \{put\} + \{Henry\} + \{Kissinger\} + \{in\} + \{his\} + \{cabinet^b\}.$$

To understand such a string one would of course have to have access to the dictionary which listed the different senses of the morphemes. How to find or contruct such a dictionary, and especially how to decide in marginal cases when two senses of a morpheme are far enough apart to assign different interpretations to them, is an unsolved problem. In the clear cases a start can be made along the foregoing lines, however, so there is hope that progress in logico-linguistics need not await a perfect understanding of the phenomenon of homonymy.

In theory, then, both semantic and syntactic ambiguities are taken care of simply by postulating that the sentence forms reaching the language automaton are fully interpreted. In practice this complicates the evidence-gathering process somewhat. It becomes necessary in the case of ambiguous sentences to make the informant aware of the particular interpretation about which he is being asked to render a judgement. If the informant understands the technical symbolism of the interpreted sentence forms, the sentence may simply be presented to him in that form; if not, the intended reading of the sentence must somehow be explained to him informally. Suppose for example an informant is asked whether the following putative implication is valid:

$$?\frac{\textit{Benjamin Franklin appreciated old men and women}}{\textit{Benjamin Franklin appreciated women}}$$

An alert informant would refuse to render a judgement until the premise sentence is interpreted for him. If he understands the bracketing conventions the investigator might make the wanted judgement clearer by presenting it to him in this way:

$$?\frac{\textit{Benjamin Franklin appreciated } [[\textit{old men}]\textit{ and } [\textit{women}]]}{\textit{Benjamin Franklin appreciated women}}.$$

With the ambiguity broken in this way the informant would presumably say "Yes, it would indeed be impossible for the credibility level of the premise to approach 1 without the credibility level of the conclusion approaching 1 too."

A language description put together according to these ideas would start out with an elaborated syntactic description specifying three things: (i) the set of all uninterpreted sentences (or 'surface structures' of sentences in the terminology of some); (ii) the set of all interpreted sentences ('deep structures', 'derivational histories', etc.); and (iii) a function from the set of interpreted sentences onto the set of uninterpreted sentences indicating which interpreted sentences are possible interpretations of which uninterpreted sentences. Parts (i) and (iii) can be left implicit if (ii) gives the interpretations in such a form that removing all special interpretive ('nonterminal') symbols from them restores the original ('surface') sentence. (This recipe for the syntactic portion of the description is only approximate, since the mapping can be many–one in the other direction when for example functionally equivalent low-level representations map into a single higher-level representation. An exact statement of the mathematical form of the syntactic rules will not be attempted here.) The remainder of the logico-linguistic description after (i)-(iii) have been specified is of the kind already discussed in earlier chapters, with the sentence set taken to be the set (ii) of all interpreted sentences.

To make the formal theory reflect these provisions for ambiguity we may define the *syntax* of a language to be a triple of form $\langle U, S, T\rangle$ where U is a non-empty set of finite sequences (representing uninterpreted sentences), S is a non-empty set of finite sequences, trees, labelled graphs, or other structures (representing interpreted sentences), and T is the ('ambiguation') function from S onto U indicating for each interpreted sentence the uninterpreted sentence it is an interpretation of. The entire theory may now be generalized along obvious lines to accommodate this notion of syntax. Thus a weighted language automaton becomes a 7-tuple $\langle U, S, T, W, Z, L, C\rangle$ where $\langle U, S, T\rangle$ is a syntax and W, Z, L, C are defined in terms of S as before.

EXAMPLE 9.1 *Completion of the Description of* $\mathfrak{L}_7$. The description of the fragment of English presented in Example 5.7 used numeric superscipts to keep track of derivational history, but their ambiguity-breaking effect is the same as if bracketings or some other device had been used. Note that the superscripts preserve only enough of the derivational structure to break syntactic ambiguities; more could have been preserved if it would have led to simplifications in the semantical rules but the need did not arise. It was assumed that there were no semantic ambiguities, but more realistically the vocabulary lists should have included, for example, separate entries $\textit{horse}^a$ and $\textit{horse}^b$ for the 'mammal' and 'jackstay' interpretations of *horse*, and similarly for some of the other words.

To make the description of the fragment complete rule A16 defining S could be relabelled 'Analyzed Sentences' and the following additional rules appended to the syntactic part of the description:

A17 (Unanalyzed Sentences)

$U = \{u |$ for some $s \in S, u$ is the result of deleting all superscripts from $s\}$.

A18 (Ambiguation Function)

$T = \{\langle s, u \rangle | s \in S$ and $u \in U$ and u is the result of deleting all superscripts from $s\}$.

The syntax of $\mathfrak{L}_7$ becomes $\langle U, S, T \rangle$. The semantical rules are unchanged. $\mathfrak{L}_7$ is the equivalence class of $\langle U, S, T, Z, L, B \rangle$ (or in the weighted generalization, $\langle U, S, T, W, Z, L, C \rangle$).

To save writing there could be no objection to leaving the rules corresponding to A17 and A18 implicit in language descriptions in which the analyzed sentences are just the unanalyzed sentences with special disambiguating symbols such as superscripts or brackets inserted. With this understanding the language description of Example 5.7 is complete as it stands, after the alphabetic superscripts are introduced to break the semantic ambiguities.

The treatment of ambiguity has been discussed here from the behavioral rather than the meaning-theoretic perspective of Chapter 6. But there is a parallel meaning-theoretic treatment and it is simple and natural: a language with ambiguities may be described by a glossary which is not a single-valued function but instead assigns in general more than one meaning to each sentence.

9.3 CONTEXT-DEPENDENCE

After all ordinary syntactic and semantic ambiguity has been dealt with there remains an especially problematic source of meaning indeterminacy described by linguists as 'deixis' and by philosophers of language as the presence of 'indexical expressions'. In general terms the problem is one of how to analyze a sentence whose meaning is dependent upon the situational and verbal context in which it is uttered. In English such context-dependence arises most obviously from the use of personal, demonstrative, or other pronouns (e.g. *I*, *those*, *it*), certain adverbials (*here*, *yesterday*), determiners in some uses (*the*, *these*), tensed verbs, and elision. Sentences with these features usually have no clear out-of-context meaning except in the presence of additional information specifying the spatio-temporal utterance situation, the speaker, the hearer(s), or other features of the circumstances of utterance.

As with other sources of ambiguity, so with deixis, a language automaton cannot be expected to cope with messages whose meaning is not clear. A step toward a solution of the problem might be to adopt something longer than the single sentence as the basic expression-type of the analysis. With some immediate verbal context available the language automaton might have some basis for such things as guessing the antecedents of pronouns. But this solution runs against the intuition that the sentence is the fundamental message-unit of language – that a sentence expresses a complete thought as we are taught in grammar school. To adopt it would be to require the language automaton to perform tasks of disambiguation which are not clearly intra-language universal, and in any case it would do nothing toward the solution of the problem of situational context-dependence. The more promising program would therefore seem to be to keep the sentence as the maximal unit of the analysis and extend the notion of an interpreted sentence still further in directions which remove any deictic elements.

To gain the perspective needed to accomplish this it is helpful to rethink the distinction made in traditional linguistics between a *sentence token* and a *sentence type*. A sentence token is an utterance event; it is the utterance of a particular sentence at a particular time, and a full description of it would have to tell all of the special circumstances attendant upon that utterance including time, place, speaker, hearer, the preceding discourse, etc. A sentence type, on the other hand, is more abstract. It is the general form of the sentence, and may be regarded as a collection of linguistically relevant characteristics held in common by all members of the class of possible tokens of the sentence. Traditionally a sentence type has been represented as a string of phonemes or other linguistic units. The ideal sentence set specified in traditional syntactic as well as the initial stages of logico-linguistic analyses is a set of sentence types.

Now, a sentence type is usually taken to be only the collection of linguistically relevant *acoustic* or articulatory features common to the class of possible tokens of the sentence. It is only the phonetic – or for written languages, graphic – patterns displayed by an utterance event, or higher-level abstractions of them, that have been deemed worthy of characterizing sentencehood. But there would seem to be no compelling reason to restict the notion of a sentence type so severely; since it is only an abstraction anyway one should feel free to define it in the most useful manner. Why not include as part of the sentence type itself indications of other linguistically relevant features of utterance events such as time and place of utterance, and so forth? Why should the patterns in the sound waves be thought to characterize the sentence type but not the time, place, transmitter, and receiver of the sound waves? In the case of what would otherwise be deictically ambiguous sentences the linguistic relevance of these other utterance features can hardly be questioned, for meaning cannot be determined without them. They are just as much linguistic features of the sentence type as the order of its phonemes.

What I am advocating is a program of token integration. The traditional notion of a sentence type is fine so far as it goes, but it needs to have integrated into it some further features conventionally regarded as characteristics of tokens only. The further features are exactly those needed to remove context-dependence and make the meaning clear: they are the particular aspects of the situational and verbal context which in fact *would* make the meaning clear to any native speaker witnessing the utterance event. Under the proposal, sentence types would have longer and possibly more complicated symbolic representations, since the situational features would have to be specified in addition to the usual sound patterns. On the other hand the class of possible particular utterance events corresponding to a sentence type will in general be smaller because of the narrower range of speech situations fitting the more highly specified sentence type. To differentiate it from the traditional sentence type let us call the result of moving closer to the token in this way a *tokenized* sentence type.

A tokenized sentence type which has also had its ordinary syntactic and semantic ambiguities broken after the fashion of the preceding section is supposed to be meaningful out of context and generally free of ambiguity of every kind. If it is not it needs to be tokenized further. But of course it goes without saying that token integration should not be carried too far.

The symbolic structures used to represent tokenized sentence types can, like the other special symbolism inserted as part of interpreted sentences, be chosen on grounds of convenience since what they denote is only indirectly observable. An obvious symbolism for indicating spatio-temporal location in a tokenized sentence type would be to specify the time and place of utterance in parentheses at the end of a traditional sentence type representation. Specification of the referents of pronouns might conveniently be inserted in parentheses after the pronouns. Speaking more generally, a tokenized sentence type may be represented by inserting the needed context features wherever appropriate in the surface structure representation or by appending it to an appropriate node in the interpreted sentence's derivation tree.

An appropriate technical symbolism for the purpose might be intricate, possibly involving some of the devices suggested by Montague and others, but an informal way of presenting tokenized sentences to informants might simply involve the insertion of appropriate explanatory phrases as in

> *I* (John Doakes) *took her* (Mrs. John Doakes) *there* (Paris) *for three weeks last summer.* (time-of-utterance: January 1975)

Even this much interpretation may be superfluous for most evidence-gathering since the informant's task of judging logical and quasi-logical relationships usually requires only that he be able to tell whether the indexical expressions involved in sentences being compared have the same referents or different – the particular things denoted do not matter. Using variables or other binding devices instead of actual names or definite descriptions, an informant should still be able to detect that an implication like the following is valid:

> *I* (x_1) *took her* (x_2) *there* (x_3) *for three weeks last summer.* (time-of-utterance: January 1974)
> ___
> *The summer before last he* (x_1) *took me* (x_2) *there* (x_3). (time-of-utterance: January 1975)

To simplify informant work still further a 'Convention of Like Interpretation' can be adopted under which the informant is given to understand that indexical expressions obviously intended to correspond in the examples put before him may be assumed to have the same denotata unless he is specifically told otherwise by the analyst. Consider for example the logical inconsistency relationship

> *I took her there last summer*
>
> *I didn't take her there last summer*
> ___

Under the Convention of Like Interpretation the two occurrences of *I* would be assumed by the informant to refer to the same person, and the referents of *her*, of *there*, and the times of utterance would also be assumed the same for the two sentences, so there is indeed a logical inconsistency. The analyst records the inconsistency as holding between any pair of a large family of tokenized sentence-type pairs, but the informant need not concern himself with the details of the system of tokenization.

EXAMPLE 9.2 *A Tensed Language*. To explore the semantics of tokenized sentence types let us consider a language $\mathcal{L}_8$ capable of expressing the three simple tenses of past, present, and future. The language is that of Example 5.4 (p. 101) except that one of the three tense markers *Pa*, *Pr*, or *Fu* always precedes each sentence. Also, to make the language more time-dependent let us reinterpret the predicate constants M, F, Y and 0 as changeable properties, say the properties of being employed, unemployed, having dependents, and having no dependents respectively. A typical (untokenized) sentence would then be '$Pa[Mb \vee \neg Yb]$' meaning that at some past point in time b was either employed or without dependents.

The tensed but untokenized sentence set S_0 of $\mathfrak{L}_8$ is $T + S$, where T is the set $\{Pa, Pr, Fu\}$ of tense markers and S is the sentence set of Example 5.4. The tokenized sentence set S^{tok} of $\mathfrak{L}_8$ is $S_0 \times \mathfrak{R}$ where $\mathfrak{R}$ is the set of all real numbers interpreted as times of utterance in some pre-established time frame. Thus $\langle Pa[Mb \vee \neg Yb], \rho\rangle$ for example is a tokenized sentence type representing the class of all tokens in which $Pa[Mb \vee \neg Yb]$ is uttered at time ρ. The model set M^{tok} needed for the structural description is the set of all functions from $\mathfrak{R}$ into the M of Example 5.4. Intuitively a member of M^{tok} is a time sequence of possible instantaneous states-of-affairs – a possible history of the employment and dependent-supporting status of a, b, and c. The satisfaction relation H^{tok} for the description of $\mathfrak{L}_8$ is specified in terms of the H of Example 5.4 by the rules that for every $s \in S, m \in M^{\text{tok}}$, and $\rho \in \mathfrak{R}$.

(i) $H^{\text{tok}}(\langle Pa \frown s, \rho\rangle, m)$ iff for some $\tau < \rho, H(s, m(\tau))$;

(ii) $H^{\text{tok}}(\langle Pr \frown s, \rho\rangle, m)$ iff $H(s, m(\rho))$;

(iii) $H^{\text{tok}}(\langle Fu \frown s, \rho\rangle, m)$ iff for some $\tau > \rho, H(s, m(\tau))$.

$\mathfrak{L}_8$ is the equivalence class containing the semantically structured language automaton with sentence set S^{tok}, model set M^{tok}, and satisfaction relation H^{tok}.

As examples of the logic of $\mathfrak{L}_8$, $\langle Pr \frown s, \rho_1\rangle$ logically implies $\langle Pa \frown s, \rho_2\rangle$ if $\rho_1 < \rho_2$, and $\langle Pr \frown s, \rho_1\rangle$ is logically inconsistent with $\langle Pr \frown \neg s, \rho_2\rangle$ iff $\rho_1 = \rho_2$. For an alternative treatment of tense logic see Montague (1974 p. 105).

9.4 LINGUISTIC INCOMPLETENESS

Consider the sentence

The king of France is bald

uttered in modern times. Bertrand Russell, to whom the sentence owes its notoriety (1905, pp. 41ff.), analyzed it as meaning that there exists one and only one king of France and that he is bald. From that analysis Russell drew the conclusion that the sentence is simply false, or at least that it can be so treated for logical purposes. Some of Russell's critics have found his analysis unconvincing where colloquial English usage is concerned, and have suggested that it is more natural to regard the sentence as neither true nor false but instead as suffering a truth-value gap. The rules of English simply fail to provide truth-conditions for the sentence when there is no king of France, according to these critics. It is not ambiguous but *meaningless* to language users informed of the nonexistence of a king of France.

Whether or not the critics are right, a satisfactory theory of language ought at least to allow their way of describing the situation to be formulated clearly and precisely. In logico-linguistic theory a gap in the rules of English takes the form of incompleteness in the language automata for English (§ 2.3). A sentence which exhibits such a gap is a sentence for which the learning operation or the credibility function has been left undefined in certain states. In the case of Russell's sentence they happen to be undefined for information states containing the information that there is no king of France. Incomplete automata are standard objects of study in atuomata theory and are included as an integral part of the logico-linguistic descriptive apparatus, so the technical means of describing language rule gaps is already at hand. When faced with an anomolous sentence such as Russell's critics claim his to be, the language analyst need only leave appropriate gaps in his statement of the rules in order to describe the phenomenon formally.

It seems plausible that many natural language phenomena might lend themselves to an analysis in terms of language automata left incomplete in certain respects. Some expressions, possibly including *the present king of France*, give rise to 'conditional incompleteness' because the credibility level of sentences containing them is defined for some states but not all. Other anomalies are cases of 'absolute incompleteness' in that the credibility level is never defined no matter what the state. Examples of absolute incompleteness would probably include

Both of the three men will be there;

That bachelor's wife is discontented;

John's height is the logarithm of minus one;

If it is raining and not raining, the Republicans will win;

and so on. The last of these examples is taken from the preceding chapter (§ 8.21), the 'Hypothesis B' examined in that chapter being in fact a serious example of a natural-language hypothesis exploiting the feature of incompleteness to obtain greater simplicity of statement and descriptive accuracy. Evidently the idea of using gaps in language rules to portray the language more faithfully is not merely academic.

In cases of absolute incompleteness it is fair to raise the question of whether the strings in question should be regarded as well-formed sentences at all. This is a matter of expediency. It does not matter whether a string is included in the sentence set with the learning and credibility functions left undefined for it, or whether it is left out of the sentence set from the start. Two language descriptions differing only in this particular should be called descriptively equivalent. In at least some cases, though, convenience is clearly on the side of the first alternative – of letting the anomalous strings be sentences in the syntactic stage of the description and then depriving them of meaning later on. Take for example *both of the three men*. If this were to be disallowed as a well-formed phrase one would have to disallow as well *both of the two plus one men*, *both of the cube root of twenty-seven men*, and so on for an infinity of ways of expressing the number three. At the same time one would have to grant well-formedness to *both of the six minus four men*, and so on. The rules of well-formedness would have to include in effect rules for all of colloquially expressible arithmetic! The more natural approach is to confine the specification of *S* to syntactic rules of the customary sort and to deal with anomolous strings later by leaving their learning and belief rules incomplete.

A troubling phenomenon of all natural languages which may or may not be closely related to incompleteness is *vagueness*. Not only are there many obviously vague expressions in a language like English (*several*, *a few*, *middle-aged*, *around noon*, etc.) but it has even been claimed with a degree of plausibility that virtually all natural-language expressions with descriptive content manifest at least a little vagueness. Like incompleteness, vagueness is not something to be cured by interpretation or disambiguation because even the speaker himself may not know precisely what he intends to convey. To put it another way, the vagueness may be part of the intended meaning.

Vagueness could well turn out to be one of the most recalcitrant problem areas in all of language analysis, and by the same token its eventual solution could be one of the most illuminating. At present it seems clear only that lines now drawn distinctly in language descriptions must somehow be blurred, possibly by the use in the metalanguage of something comparable to the fuzzy set theory proposed by Lotfi Zadeh (1965). But it is far from clear *which* distinctions should be blurred. Should the completeness/incompleteness dichotomy be made a matter of degree? Or should the fuzziness be introduced someplace else such as the learning and credibility functions, the satisfaction relation, or the denotation mappings within the modelling structures? The behavioral correlates of vagueness, without which no systematic evidence on the subject can be gathered, are still unclear. There may even be something inherently self-defeating in an attempt to clarify vagueness; in any case the attempt will not be made here.

9.5 NON-DECLARATIVE SENTENCES

How are interrogative, imperative, exclamatory, and other non-indicative sentence types to be treated in a logico-linguistic description? The solution to this problem is, I believe, to recognize that there is no problem. There is no need to alter the foundations of the theory in any fundamental way on account of non-declarative moods for the reason that logico-linguistics has to do with an aspect of linguistic communication which is largely independent of mood.

The aspect of utterances with which logico-linguistics is concerned is their informative aspect. It treats only the conveyance of information from speaker to hearer, not what the hearer will then say or do in response to the information nor what the speaker may hope or intend that the hearer may say or do. The Input and Output Selectors are left out of the analysis along with the entire Nonverbal Subsystem. To the only behavioral entity that *is* of concern – the Language Automaton – declarative, interrogative, and imperative statements are all alike; they are all just state-reporting and state-changing signals.

Consider for instance a yes–no question such as

Is John at home?

This sentence conveys information like any other – specifically, the information that the speaker of the sentence desires to know whether John is at home. An approximate paraphrase might be

I desire to know whether John is at home.

The difference between this declarative paraphrase and the interrogative original is significant not for the Language Automaton but for the Output Selector. It is important to the speaker's output choice mechanism because the speaker knows the interrogative form will probably elicit a 'yes' or 'no' response from the hearer, a conventionalized response which transmits the desired information with extraordinary efficiency, whereas the declarative form might not elicit such a useful response. And the difference is important to the hearer's output choice mechanism because the hearer will tend to give one kind of response to the interrogative form and another to the declarative. Thus interrogativeness would be an important property to study in higher pragmatics but on the logico-linguistic level the interrogative mood has much the same status as the prefactory phrase *I want to know whether* ______ .

As a consequence of the indifference to interrogativeness, logical and quasi-logical relationships apply to questions as much as to any other sentence type. For example,

Is John at home in bed?
————————————
Is John in bed at home?

should probably be counted as a valid logical implication in English. If the speaker believes he wants to know whether John is at home in bed, then he must also believe he wants to know whether John is in bed at home. (It is assumed by the Convention of Like Interpretation that John, the utterer, and the time of utterance are the same for both sentences.) This does not mean, however, that the logic of questions is a carbon copy of the logic of statements. Negating the main verb of a question doesn't necessarily produce an opposite meaning, for instance, as is seen in this example:

Is John at home?

Isn't John at home?
* ————————

Far from being logically contradictory, a case could even be made for the logical equivalence of these two questions. Both mean approximately, "I need information about whether John is at home", the fact that one would answer 'Yes' to the first when 'No' would be appropriate to the second being an observation about the input and output selection rules of English rather than its language automaton.

The situation with respect to commands, exclamations, and other non-declarative sentence types is similar to that of questions and need not be discussed separately. There is however a special class of declaratives which are in some ways less like ordinary declaratives than any of the non-declarative types. They are the assertions called 'performatives', examples of which are

I promise to return the book on Thursday;

I now pronounce you man and wife.

As performatives these sentences have the unusual property that under appropriate circumstances the mere uttering of them makes them true. Because of this the Output Selector need not consult the Language Automaton at all about the current credibility level of such sentences before deciding to utter them. Instead, their high credibility weight after utterance is a *result* of their utterance. But this reversal of cause–effect relationships which characterizes performatives is part of higher pragmatics and has nothing to do with a sentence's logico-linguistic properties. As with ordinary declaratives one has, for example, such logical properties as the following inconsistency relationship:

I promise to return the book by Thursday

I don't promise to return the book by Thursday

9.6 Physical Realizability

By a well-known metamathematical result (Church's Theorem), there is no mechanical decision procedure for typical first order predicate logics of the classical sort. From this it follows (under Church's Thesis) that it would be impossible to obtain a physical realization of a language automaton for such a language. There is no hope, for example, of programming a digital computer to act as the automaton. Similar remarks apply to any other languages whose expressive power is at least equal to that of such languages, and it has been conjectured that this probably includes all the natural languages. If this conjecture is correct, *most of the more interesting languages probably do not have physically realizable structuralizations.*

It might seem paradoxical at first that a language spoken by humans might not be physically realizable, not even in the human brain. It has to be remembered, though, that the rules of language about which this assertion is made are highly idealized. No actual speaker-hearer has complete mastery of them; all that is claimed for them is that they are the rules the language users *try* to abide by as consistently as they can. In an exclusively syntactic description rules are commonly stated which allow for sentences so long it would be physically impossible to utter them, yet these are said to be grammatical in the sense that an *ideal* speaker-hearer would consider them so. And in a logico-linguistic description, sentences are in addition assumed either believed or not believed in a state even though they may be so long and logically complex that only an ideal intellect could decide whether to believe them or not. With such strong idealizations already in force it should be clear that no natural language automaton is ever going to have a perfect physical realization anyway, even if it were not for the problem of theoretical undecidability. Little is added by the discovery that a realization is impossible even in principle.

How do the language users manage to get along at all, then? What saves them is the fact that even when a language is theoretically undecidable it can have decidable sublanguages so inclusive as to be virtually indistinguishable from the whole language for practical purposes. In the classical predicate logic, for instance, decision procedures have been discovered which apply to so much of the calculus that one must know what one is doing even to be able to find sentences that lie outside the testable subdomain. Thus by restricting the classical predicate languages in ways which would hardly be noticed for the practical purpose of exchanging factual information (as a last resort, by putting a limit on sentence length), one could render them decidable. It seems likely that a similar situation obtains for the natural languages. If so, the spectre of theoretical undecidability is a bogeyman. One should not care about undecidability at all if an abstract language description is all that is wanted, whereas if for some practical reason the description is to be implemented on a computer there will in all probability be ways to get around the problem.

Realizations for the more interesting languages may be impossible, then, but 'virtual realizations' good enough for all practical purposes are probably possible. Programming a digital computer as a virtual realization is simply a matter of finding decision rules of sufficient efficiency and coverage to serve the purpose for which the realization is needed. For complicated languages this may be practically difficult but it is not theoretically impossible. The possibility of obtaining computer realizations or virtual realizations of language structuralizations has some interesting implications for research in language-oriented fields of computer application, a few of which will be explored in the remainder of the chapter.

To make the concept of physical realizability clearer it can be defined in terms of certain types of automata corresponding to actual computing devices. Clearly any finite automaton is in principle realizable since all that would be required to realize it would be an appropriately programmed digital computer of conventional design and sufficient internal memory capacity. In addition to finite automata there is an important class of automata called 'growing automata' (Burks and Wang 1957) which are $\langle X, Y\rangle$-finite and Z-denumerable but which are physically realizable in a sense in spite of their potential infinity of states. They are devices whose memories, though finite at any given moment, are indefinitely extendable; examples include Turing machines and conventional digital computers with external tape units.

A language automaton can now be defined as *realizable* if and only if it can be simulated by a finite or growing automaton. For the precise definition of 'simulated' the reader is referred to the concept of weak simulation defined by Arbib (1969). Here we point out only that it is often feasible to simulate automata that are not X-finite by finite or growing automata that are. Thus even though the sentence set of a language automaton is normally infinite, it may be possible to simulate it anyway with a finite or growing automaton by accepting the input sentences one symbol at a time – phoneme by phoneme, say, instead of sentence by sentence.

9.7 AUTOMATIC QUESTION-ANSWERING

Suppose one had a computer realization or virtual realization of a language automaton for some reasonably expressive language. To what use might it be put? By first feeding a large corpus of factual sentences into it as learning input, one could use it as what information retrieval specialists call a 'question-answering' or 'fact retrieval' system. Queries may be submitted to a language automaton in which information has been stored in this way in the form of test sentences, and the automaton's responses are its answers to the queries. There is a class of question-answering systems, in other words, which are essentially just realizations of their own input languages.

To see in more detail how such a system might work, suppose a computer were programmed to act the part of a binary-weighted language automaton for some semantically structuralizable language containing a sentence operator comparable to classical negation. It would be possible to exploit such a program to answer yes–no questions in that language – or what comes to the same thing, True–False questions. As a preparatory step the system would be set in a nescient state corresponding to the empty store of information. The information to be stored, expressed as a series of sentences of the language, would be entered next as learning input. Submitting a True–False query to the system would be accomplished by submitting the query sentence as test input followed by the negation of the sentence. If the unnegated sentence produces the 'Yes' ('Believed') response, the system is regarded as having given the answer 'TRUE'. If instead the negation of the sentence yields a 'Yes' response this is interpreted as the answer 'FALSE'. If neither the sentence nor its negation produces a 'Yes' response the stored information is insufficient to answer the question and the system is regarded as having answered 'DON'T KNOW'.

One of the earliest experimental question-answering systems possessing genuine deductive power worked in precisely this way (Cooper 1964, 1965). The language for which the computer was programmed to act as a structuralization was a fragment of English adequate for expressing simple taxonomical information. Given a vocabulary taken from an elementary chemistry textbook, the language contained such sentences as *Sulfur burns* and *All solid metalic magnesium compounds which are not brittle are white compounds that burn rapidly*. With the sentences *Gasoline is a fuel*, *Gasoline is combustible*, and *Combustible things burn* among the data-sentences entered into the system, it was able for example to answer 'TRUE' to the query-sentence *Gasoline is a fuel that burns*. The logico-linguistic problems involved in the analysis of this particular fragment of English were of course relatively trivial. Practical True–False question-answering systems await the analysis of more substantial sublanguages of English and more efficient computer realizations of them.

Other types of question-answering systems are obtainable from weighted language automata having more than two credibility weights. For instance, a language automaton with the three weights 'Believed', 'Disbelieved', and 'Undecided' would be able to give True–False answers with only one test instead of two, and in languages not having any device corresponding to classical negation. The most interesting prospect is that of a probability-weighted language automaton which, if it could be realized, would be able to give probability estimates rather than True–False answers. The probability-weighted case is of special interest not only because probabilistic answers are more informative than True–False answers but also because probability-weighted language automata are not intransigent: their stores of information can be changed or corrected simply by entering the new information which takes precedence as learning input. The specification of an appropriate nescient state in which to initialize a probabilistic question-answering system is one of the unsolved problems connected with this kind of system.

Systems able to answer certain other kinds of specific questions – who-, when-, and where-questions, for example – can be based on algorithms making repeated use of language automaton realizations. For example, in a question-answering system about American history the question "Who was president of the United States in 1930?" could in principle be answered by successive substitution of all proper names in the input language into the frame ______ *was president of the United States in 1930* until a name was found which when inserted in the frame yielded a high credibility weight as a test sentence. Admittedly this approach solves the problem only when proper-name answers are what is wanted. Also, it is probably not the most efficient way to go about answering who-questions. However, more efficient strategies would still require as a linguistic basis something comparable to a logico-linguistic analysis of the input language.

Even for question-answering systems which do not have the exact form of a language automaton, then, it might still be appropriate to encapsulate the results of the associated linguistic and logical research in the form of a logico-linguistic description of the proposed input language. Such a description could serve as the linguistic component of the research with the help of which a number of different kinds of question-answering systems could be designed by superimposing appropriate searching and answering strategies.

9.8 ENTHYMEMES, ANALYTICITY

In traditional logic an argument with a suppressed premise is an 'enthymeme'. For example, the invalid argument

Harry Truman often campaigned in his home state.
* ———————————————
Harry Truman often campaigned in Missouri.

becomes valid when regarded as an enthymeme whose hidden premise is the one indicated within brackets in

Harry Truman often campaigned in his home state.
(*Harry Truman's home state was Missouri.*)
———————————————
Harry Truman often campaigned in Missouri.

Some have conjectured that most of the arguments encountered in casual discourse are enthymemes – that in informal reasoning unstated premises are more the rule than the exception.

In the traditional account of enthymemes it is not always quite clear just what a 'suppressed premise' is supposed to be like. Fortunately, in logico-linguistic terms enthymemes are explainable without having to resort to the awkward fiction of suppressed premises at all. It is not suppressed premises that make an enthymeme but assumed information. The foregoing example is a valid argument in the sense that in every information state in which Harry Truman's home state is known to be Missouri, if the explicit premise is believed so is the conclusion. Doing away with hidden premises in favor of restrictions on the state set in this way circumvents what would otherwise be awkward questions about the exact form of the hidden premise, about how there could be such a premise in languages not rich enough to express it, and so on.

From a more practical point of view, enthymemes are something to be guarded against when employing informant technique. The danger is that a careless informant might unconsciously make use in his thinking of an unexpressed factural assumption without telling the analyst he was doing so. Informants should be warned of this danger and possibly even be given practice in detecting hidden assumptions. When confronted with an enthymeme the informant must be able to recognize that it is invalid as it stands. He may however wish to tell the analyst about the additional information that would make it valid in case the analyst might be interested in extending the argument by adding another premise.

There is no theoretical difficulty in any of this until an unexpressed factual assumption is encountered which borders on being a logical truism. In such marginal cases it may be difficult to know whether to call the corresponding additional premise a tautology, in which case the argument is valid without it, or non-tautologous, in which case the argument is an enthymeme and invalid without it. An example would be:

? *That ball is red all over*
——————————— .
That ball is not blue all over

Is this argument valid?

To be valid it would have to invoke an assumption to the effect that nothing red all over is at the same time blue all over. But philosophers have been unable to agree whether

Nothing is both red all over and blue all over

is tautologous or not, the sentence in question having in fact been the object of a protracted controversy in the journal literature. So one cannot really tell whether one is dealing with an enthymeme or not without having first settled upon a precise and uniformly applicable criterion of logical validity or tautologousness. (The problem is sometimes discussed in terms of defining 'analyticity' seen as a weaker property than logical validity, but we will not need to introduce this fine distinction here.)

The analytic–synthetic dichotomy, which is what is at issue in the problem of classifying premises as tautologous or not, has been a prolific source of debate since at least the time of Kant. Is the sentence

? ———————————
All cars are vehicles

best treated as analytic or synthetic? The related distinction with respect to logical implication is also troublesome; is

$$? \frac{\textit{That is a car}}{\textit{That is a vehicle}}$$

a logical implication or only a material one? And is

$$? \frac{\textit{That is a car}}{\textit{That is not a vehicle}}$$

logically inconsistent or only factually so? In the unifying perspective of logico-linguistic theory all three questions, plus others in the same family, can be replaced by the single question of whether information states can exist in which cars are not known to be vehicles. More generally, the whole complex of issues surrounding the analytic-synthetic distinction can be summed up in logico-linguistic terms as the issue of what states to exclude from the state set.

With the issue put in that way it seems natural to distinguish between *logically strong* and *logically weak* language descriptions. A logically strong description of a language is one which tends to disallow questionable information states, with the result that many tautologies, implications, and inconsistencies exist among the sentences of the language. A logically weak description allows more information states of the marginal sort in the state set, producing fewer logical relationships and more logical independence. The traditional philosophical problem of the analytic–synthetic dichotomy therefore presents itself to the working analyst as the policy decision of whether to make his language description logically strong or logically weak, and beyond that of just where to draw the boundary line. (A more radical option would be to abandon the assumption that there is a sharp boundary, generalizing the entire theory along lines which make membership in the state set a matter of degree, or possibly of degree and kind. This approach would reflect well the anti-empiricist philosophy of W.V.O. Quine (1951).)

The philosophical problem of how to state a precise criterion of analyticity is a perennial one which we will not attempt to resolve here. My own view of the issue is simple-minded: I suspect the problem has remained unsettled for so long mainly because it has been underspecified. Suppose an architect were to stroll by a construction site and a carpenter about to saw the end off a plank were to ask him, "How long should I make this board?" The problem is clearly underdetermined and the only sensible reply would be for the architect to ask in turn, "What do you need it for?" Note that if the carpenter responds, "I am building a house on the foundations you see over there", the problem is *still* underspecified; the only way for the architect really to help the carpenter would be for them to go into great detail about just what the board is needed for, what the alternative uses for it might be, what kind of a house would be best, and so forth. The decision is a design decision, it is by nature complex and because of the uncertainties involved there may not be any one right answer.

Analogously an analyst investigating a natural language might ask a language theorist, "How logically strong should I make my language description?" To this the theorist can only reply, "What do you want it for?" The answer "I want a logico-linguistic analysis based on such-and-such mathematical foundations" is a good start but still leaves the problem vastly underdetermined. Is the description to be used as an aid in teaching the language to foreigners, as a starting point from which to work toward an eventual psychological performance model, or what? There is no reason to suppose that all potential uses to which a language description could be put call for the same logical strength. Hence without knowing the intended purpose of the description the question simply cannot be answered, and even after the purpose is specified the matter may still call for a delicate design decision.

Continuing the example, suppose the analyst is able to say definitely, "The language description I am working on is to be used as a basis for programming a question-answering system", and to tell the topic area, the computer it is to be implemented on, and many other details. This statement of purpose is specific enough so that the beginnings of an answer *can* be given, or at least a clarification of the kinds of considerations that would lead to an answer. For a question-answering system, a logically strong description will probably require more work on the analyst's part and a longer description, and probably more work on the programmer's part in implementing the description on a computer; but there will afterward be less need to enter into the system as learning input large numbers of factually trivial statements such as *All cars are vehicles*. A logically weak description, on the other hand, will be easier to write and implement but will necessitate more trivial input. Clearly this is a design trade-off to be decided on grounds of practicality, for the system will end up responding to questions with the same answers in either case.

An interesting aspect of this design decision is that whichever approach is taken the language description must be at least logically strong enough to get the question-answering system to the 'take-off point' at which further trivial information can be entered into the working system in the form of sentences of the language. Since this sort of situation is likely to recur in other potential computer applications, the 'take-off point' suggests itself as a minimum logical strength which an analyst ought to attempt to build into any serious general-purpose language description. In a description of English, one would probably have to give a logically strong treatment of at least the 'function' words and constructions (*all*, *and*, *if*, *is*, *of*, *the*, subject–predicate constructions, quantifiers, etc.) to get to the take-off point. However the logical strength of the treatment of many descriptive words could be left to the discretion of the analyst under this rule (e.g. *car* and *vehicle* need not be logically related in the description because of the availability of *All cars are vehicles* as an input sentence). This leaves much that is arbitrary about the dividing line between analytic and synthetic. But our argument is that a certain amount of arbitrariness is unavoidable in a general purpose description because such a description has by definition no unique intended purpose.

We have seen that the problem of deciding on logical strength can be viewed more fundamentally as the question of what sorts of information states should be allowed to exist in the state set. The answer to the latter question can never be derived from logico-linguistic theory proper because it is essentially a question about the interpretation of 'information state', a primitive and hence undefinable term of the theory. Thus the theory can never resolve the philosophical controversy about analyticity and should not be expected to do so. However, it does pose the problem in the new light of a question about the existence of certain problematic information states. And, for the practically minded, it is flexible enough to allow the analytic–synthetic line to be drawn wherever it might most helpfully be assumed to lie for the application at hand.

9.9 FURTHER COMPUTER APPLICATIONS

It has already been pointed out that a (virtual) realization of a structuralization for a language can serve immediately as a question-answering system in that language. The question-answering application seems appropriate for languages which, like the natural languages, are rich enough to express a wide variety of factual information. But suppose instead a (virtual) realization were available for a mathematician's or logician's formal language – a language whose strong point is not richness of factual expression but clarity, conciseness, and freedom from ambiguity. What would it be good for? Such a realization would be best regarded as a theorem-proving system rather than a fact retrieval system. To use it for theorem-proving (more precisely, as a decision device) one would set it in a nescient state, enter the hypotheses of the conjectured theorem as learning input, enter the conclusion as a test input, and observe whether the 'Yes' or 'No' output was forthcoming. Provided the language in question is semantically structuralizable, a 'Yes' output would indicate theoremhood (by Theorem 5.4 and the Corollary of Theorem 5.2).

Research done on automatic theorem-proving may therefore be regarded from the present perspective as the study of how to obtain efficient realizations for structuralizations of certain formal languages. Conversely logico-linguistics provides a general metatheory for theorem-proving research – a metatheory by no means complete in itself, but which with the addition of efficiency considerations might be helpful. Similar remarks apply to question-answering. The efficiency criteria are of course different for theorem-proving than for question-answering: if a realization is to be used for theorem-proving it must be good at handling a few logically intricate sentences, whereas if it is to used for fact retrieval it must be able to handle a large corpus of relatively simple sentences. But in principle theorem-proving and yes–no question answering are much the same.

There are other potential computer applications for which logico-linguistics, though not the whole underlying theory, is nonetheless a necessary part. The mechanical translation of natural languages is an example. To obtain a precise theory of translation one might take as a starting point the idea of a 'bilingual language automaton' – a language automaton whose sentence set includes the sentences of both source and target languages, but which is otherwise like an ordinary language automaton. A bilingual language automaton is a model of the linguistic competence of a bilingual speaker. The notion of a bilingual language automaton immediately yields a theory of exact translation *via* the logico-linguistic definition of synonymy (§ 6.4): a sentence in the target language is an exact translation of a sentence in the source language if and only if the two are synonymous (behaviorally indistinguishable) in the bilingual automaton.

A theory of exact translation is not of itself a sufficient basis for mechanical translation because, among other reasons, an 'approximate' translation is wanted whenever no exact translation exists. Reasonable criteria for approximate translation are not easy to state, and the theory of approximate translation may not lie entirely within the scope of logico-linguistics proper. However, machine translation is unlikely to advance much beyond its present plateau until these theoretical problems are faced, and the logico-linguistic framework at least suggests a possible starting point for the development of the needed theory.

The problem of how to resolve ambiguities comes up in all computer applications in which natural language input must be handled 'understandingly'. This includes mechanical translation, natural-language question-answering, certain other types of information retrieval, speech and handwriting recognition, and the typewriter problem (voice-to-print translation). Strategies for breaking ambiguities do not properly belong under the heading of logico-linguistics, for disambiguation is part of higher pragmatics as explained earlier. However, the Disambiguator must normally make use of the Language Automaton by submitting candidate interpretations to it to find out their current credibility level. A language automaton does not disambiguate, then, but it is nevertheless essential to have access to a language automaton in order to disambiguate well. This alone makes logico-linguistics a necessary *part* of the theoretical apparatus needed for almost all the more advanced language-processing applications.

To illustrate the role of the language automaton in the resolution of ambiguity, consider the sentence

The box was in the pen.

This particular sentence is of historic interest as the sentence on which Bar-Hillel based his argument for the impossibility of fully automatic high quality mechanical translation (1964 p. 175). The problem is of course to decide whether to interpret the sentence (in the notation of § 9.2) as

$$\{The\} + \{box\} + \{was\} + \{in\} + \{the\} + \{pen^a\}$$

or

$$\{The\} + \{box\} + \{was\} + \{in\} + \{the\} + \{pen^b\},$$

where $\{pen^a\}$ is the morpheme for the writing instrument and $\{pen^b\}$ for the enclosure. Even examining the sentence out of context, an English speaker can tell that the $\{pen^b\}$ interpretation was probably intended, so it is factual information about the real world that provides the deciding clue as Bar-Hillel pointed out. Indeed if the sentence occurred in a context in which *pen* had already been used in both senses, the hearer would have no other grounds on which to disambiguate it. In logico-linguistic terms, it is as though the hearer were to submit both possible interpretations to his Language Automaton and choose the one with the higher credibility weight as the preferred interpretation.

If a computer realization of a language automaton were available and could be set into a reasonably typical information state incorporating a large body of trivial worldly information, there is no reason why this same criterion could not be applied to resolve the ambiguity by computer. A well-informed realization of a language automaton is therefore an important key to disambiguation of human quality. To store enough information in a language automaton to be able to resolve all or most humanly resolvable ambiguities would indeed be a formidable task, and perhaps it will never be achieved. But never is a long time, and since the disambiguation capability is needed for so many important applications it would be foolhardy to predict that it will not eventually be done.

9.10 ARTIFICIAL INTELLIGENCE

Finally, logico-linguistics has a unique place in the quest for artificial intelligence. It is interesting that Turing's famous criterion for telling whether a machine can think is stated entirely in terms of verbal behavior. By Turing's test a computer is said to 'think' if its verbal intercourse is indistinguishable from that of a human. Other criteria that have been proposed for artificially intelligent behavior are broader, demanding in addition to humanlike verbal behavior a capacity for pattern recognition, game playing, execution of delicate physical manipulations, and so forth. But a sophisticated linguistic capacity has been considered basic to artificial intelligence by all commentators, and this is reason enough to suppose that a logico-linguistic analysis or something comparable must be implicit in the design specifications of any artificially intelligent device.

The role of logico-linguistics in machine intelligence can be pinpointed more precisely in terms of the diagram of Figure 2.4 (p. 34). To pass Turing's test, a device must be designed which is a realization of the Verbal Subsystem of that diagram. Clearly, then, even the basic capacity for humanlike verbal intercourse goes beyond mere logico-linguistic considerations, embracing as it does an ability to do intelligent input and output selection including interpretation and other higher pragmatic tasks. The Language Automaton is nonetheless an indispensable *component* of the Verbal Subsystem, making of logico-linguistics an essential part, though by no means all, of the theoretical basis for artificial intelligence under Turing's criterion.

To go beyond Turing's criterion and achieve certain kinds of intelligent nonverbal input and output as well, more than just the Verbal Subsystem has to be realized. Redrawing Figure 2.5 (p. 36) in the manner of Figure 9.1, we may schematize the needed additional components as a 'Nonverbal Input Selector' which analyzes nonverbal sensory input, and a 'Nonverbal Output Selector' which decides upon the nonverbal motor output. The information automaton containing the language automaton as a subautomaton emerges as the pivotal component of the entire system. It contains all of the device's current factual beliefs, whether linguistically expressible or not. This belief state may be altered at any time either by sentences, interpreted and assigned a

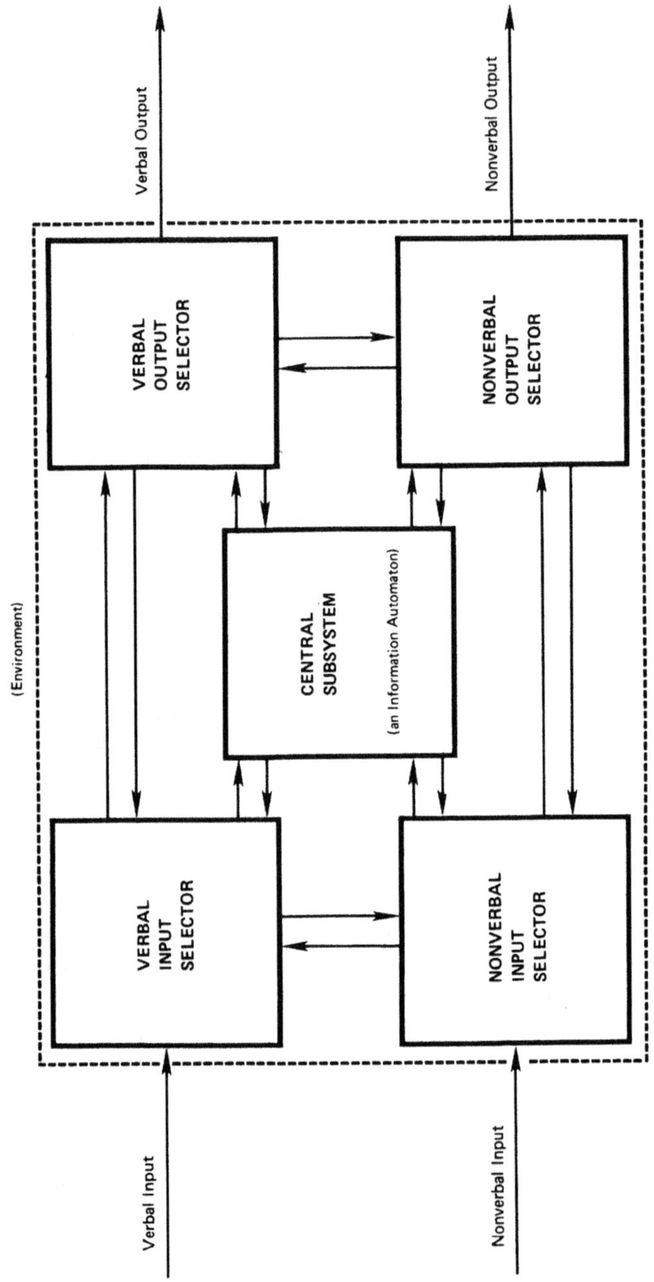

Figure 9.1. Possible configuration of components for an artificially intelligent device.

credibility weight by the Verbal Input Selector, or by sights, sounds, and other sensory data interpreted and predigested as learning input signals of an as yet unknown form by the Nonverbal Input Selector. The state of the information automaton is consulted as part of the decision process for what to say by the Verbal Output Selector, and for what to do by the Nonverbal Output Selector. The information automaton is in a manner of speaking the 'knowing' element of the artificially intelligent device. It is not wholly describable in logico-linguistic terms, but it is partly so. If one had a precise description of this information automaton, a structuralization of the language would result from a restriction of its input–output to verbal signals.

An artificially intelligent device is supposed to make intelligent decisions, both about what to say and what to do. If in such a device the information automaton is probability-weighted and semantically structuralizable, the place of utility theory and decision theory in the decision-making process becomes apparent. In a semantically structured structuralization of the information automaton there is at any given moment a probability measure over the model set indicating what is currently known. To this probability measure may be added a utility measure indicating (up to a linear transformation) the current worth or 'desirability' (Jeffrey 1965) of the various possible events to the device. Taken together, two such measures are known to be sufficient to specify rational decisions for action under familiar Bayesian formulae for maximizing expected utility. This is a sketchy but perhaps at least suggestive account of how by adding enough additional capabilities to a language automaton one could eventually end up with a Bayesian robot. The point, again, is that a logico-linguistic analysis is an integral part of what is needed.

9.11 THE FUTURE

One need have no illusions that an accurate logico-linguistic description of an entire natural language can be written overnight. If even the syntactic analysis of English is not yet fully complete, and this in spite of many years of effort by talented linguists, how much more effort is a full-fledged logico-linguistic analysis likely to require? However, it is possible on the logico-linguistic level as it has been on the syntactic to make a start by analyzing significant language fragments and addressing the theoretical problem areas one at a time. Investigations of specific linguistic phenomena are of interest in themselves and their cumulative effect is to bring the goal of a full analysis ever nearer. This sort of logico-linguistic research program, or something like it, must eventually be undertaken if man the communicator-ratiocinator is ever to be fully understood. If scientific curiosity does not impel the research first, ultimately the computer applications will.

REFERENCES

Adams, E.W. 1965. 'The Logic of Conditionals', *Inquiry* **8**: 166-97.

–, 1966. 'Probability and the Logic of Conditionals', in *Aspects of Inductive Logic*, ed. J. Hintikka and P. Suppes, pp. 265-316. Amsterdam: North-Holland.

–, 1970. 'Subjunctive and Indicative Conditionals', *Foundations of Language* **6**: 89-94.

–, 1975. *The Logic of Conditionals: An Application of Probability to Deductive Logic*. Dordrecht: D. Reidel.

Anderson, A.R. and Belnap, N.D. 1962. 'The Pure Calculus of Entailment', *Journal of Symbolic Logic* **27**: 19-52.

Arbib, M.A. 1969. *Theories of Abstract Automata*. Englewood Cliffs: Prentice Hall.

Bar-Hillel, Y. 1964. *Language and Information*. Reading: Addison-Wesley.

–, 1970. *Aspects of Language*. Amsterdam: North-Holland.

Birkhoff, G. 1948. *Lattice Theory*. Revised edition. *American Mathematical Society Colloquium Publications*, vol. 25. New York.

Bloomfield, L. 1933. *Language*. New York: Holt, Rinehart and Winston.

Bryant, E.C. 1966. *Statistical Analysis*. 2nd ed. New York: McGraw-Hill.

Burks, A.W. 1951. 'The Logic of Causal Propositions', *Mind* **60**: 363-82.

Burks, A.W. and Wang, H. 1957. 'The Logic of Automata', *J. of the Association for Computing Machinery* **4**: 193-218 & 279-97.

Carnap, R. 1942. *Introduction to Semantics. Studies in Semantics* vol. 1. Cambridge, Mass: Harvard University Press.

–, 1963. *The Logical Foundations of Probability*. 2nd ed. Chicago: University of Chicago Press.

Carnap, R. and Jeffrey, R.C., eds. 1971. *Studies in Inductive Logic and Probability*, vol. 1. Los Angeles: University of California Press.

Carroll, Lewis. 1894. A Logical Paradox. *Mind* **3**: 436-8.

Chisholm, R.M. 1946. 'The Contrary-to-Fact Conditional', *Mind* **55**: 289-307.

Chomsky, N. 1955. 'Logical Syntax and Semantics: Their Linguistic Relevance', *Language* **31**: 36-45.

–, 1965. *Aspects of the Theory of Syntax*. Cambridge, Mass: M.I.T. Press.

Church, A. 1956. *Introduction to Mathematical Logic*, vol. 1. Princeton: Princeton University Press.

Cohen, J. 1960. *Chance, Skill and Luck: The Psychology of Guessing and Gambling*. Baltimore: Penguin Books.

Cooper, W.S. 1964. *Set Theory and Syntactic Description*. The Hague: Mouton.

–, 1964. 'Fact Retrieval and Deductive Question-Answering Systems', *J. of Association for Computing Machinery* **11**: 117-37.

–, 1965. 'Automatic Fact Retrieval', *Science Journal* **1**: 81-6 (no. 4).

–, 1968. 'The Propositional Logic of Ordinary Discourse', *Inquiry* **11**: 295-320.

–, 1976. *The Indefinite Article of English: A Case Study in Logico-Linguistics.* Xeroxed. School of Library and Information Studies, University of California, Berkeley.

–, 1978. 'The Logico-Linguistic Evidence Underlying Montague's Language Descriptions', *Synthese* **38**.

Edwards, W., Lindman, H., and Savage, L. 1963. 'Bayesian Statistical Inference for Psychological Research', *Psychological Review* **70**: 193-242.

Ellis, B. 1973. 'The Logic of Subjective Probability', *British Journal for the Philosophy of Science* **24**: 125-52.

–, (Forthcoming.) *Epistemological Foundations of Logic.*

Empiricus, Sextus (translation) 1935. *Against the Logicians*, book 2, R.G. Bury, trans. Cambridge, Mass: Harvard University Press.

Faris, J.A. 1962. *Truth-Functional Logic.* New York: Free Press of Glencoe.

Frege, G. 1879. 'Begriffschrift', translated in *Translations from the Philosophical Writings of Gottlob Frege*, 2nd ed., 1960, eds. P. Geach and M. Black, pp. 1-20. Oxford: Basil Blackwell.

Gentzen, G. 1935. 'Untersuchungen über das logische Schliessen', *Mathematische Zeitschrift* **39**: 176-210 and 405-31.

Ginsburg, S. 1962. *An Introduction to Mathematical Machine Theory.* Reading: Addison Wesley.

Hilbert, D. and Bernays, P. 1934. *Grundlagen der Mathematik.* Berlin: Springer.

Jeffrey, R.C. 1964. 'If' (abstract), *Journal of Philosophy* **61**: 702-3.

–, 1965. *The Logic of Decision.* New York: McGraw-Hill.

Johnson, D.A. 1963. *Logic and Reasoning in Mathematics.* St. Louis: Webster.

Kalish, D. and Montague, R. 1964. *Logic: Techniques of Formal Reasoning.* New York: Harcourt, Brace and World.

Kaplan, A. 1964. *The Conduct of Inquiry: Methodology for Behavioral Science.* San Francisco: Chandler.

Kleene, S.C. 1967. *Mathematical Logic.* New York: John Wiley.

Kripke, S.A. 1959a. 'A Completeness Theorem in Modal Logic', *J. of Symbolic Logic* **24**: 1-15.

–, 1959b. 'Semantical Analysis of Modal Logic' (abstract), *Journal of Symbolic Logic* **24**: 323-4.

–, 1965. 'Semantical Analysis of Intuitionistic Logic I', in *Formal Systems and Recursive Functions*, eds. J.N. Crossley, M.A.E. Dummett, pp. 92-130. Amsterdam: North-Holland.

Lewis, C.I. 1918. *A Survey of Symbolic Logic.* Berkeley: University of California Press.

Lewis, C.I. and Langford, C.H. 1932. *Symbolic Logic.* New York: The Century Company.

Lewis, D. 1976. 'Probabilities of Conditionals and Conditional Probabilities', *Philosophical Review* **85**: 297-315.

Lukasiewicz, J. 1951. *Aristotle's Syllogistic from the Standpoint of Modern Formal Logic.* Oxford: Clarendon Press.

MacKay, D.M. 1969. *Information, Mechanism and Meaning.* Cambridge, Mass: M.I.T. Press.

Martin, R.M. 1959. *Toward a Systematic Pragmatics.* Amsterdam: North-Holland.

Montague, R. 1974. *Formal Philosophy: Selected Papers of Richard Montague*, ed. R.H. Thomason, New Haven: Yale University Press.

Moore, E.F. 1956. 'Gedanken Experiments on Sequential Machines', in *Automata*

Studies, Ann. Mathematical Studies vol. 34. eds. C.E. Shannon and J. McCarthy, pp. 129-153. Princeton: Princeton University Press.

Partee, B.H., ed. 1976. *Montague Grammar.* New York: Academic Press.

Peirce, C.S. 1885. In *Collected Papers of Charles Sanders Peirce* vol. 3, eds. C. Hartshorne and P. Weiss, 1933. Cambridge: Harvard University Press.

Polya, G. 1954. *Patterns of Plausible Inference, Mathematics and Plausible Reasoning* vol. 2. Princeton: Princeton University Press.

Popper, K.R. 1959. *The Logic of Scientific Discovery.* London: Hutchinson.

Quine, W.V.O. 1941. *Elementary Logic.* New York: Ginn and Company.

–, 1953. 'Two Dogmas of Empiricism', in *From a Logical Point of View,* ed. W.V.O. Quine, Cambridge, Mass: Harvard University Press.

Ramsey, F.P. 1931. *The Foundations of Mathematics and Other Logical Essays.* New York: Harcourt Brace.

Rescher, N. 1962. 'Quasi-Truth Functional Systems of Propositional Logic', *J. of Symbolic Logic* **27**: 1-10.

Rogers, R. 1963. 'A Survey of Formal Semantics', *Synthese* **25**: 17-56.

Russell, B. 1905. 'On Denoting', *Mind* **14**: 479-93. Reprinted with commentary in *Contemporary Readings in Logical Theory,* eds. I.M..Copi and J.A. Gould, 1967, pp. 93-105. New York: MacMillan.

Saussure, F. de. 1916. *Course of General Linguistics.* Translated by W. Baskin, 1959. New York: Philosophical Library.

Scott, D. and Strachey, C. 1971. 'Toward a Mathematical Semantics for Computer Languages', in *Computers and Automata,* ed. J. Fox, pp. 19-46. New York: John Wiley.

Simons, L. 1965. 'Intuition and Implication', *Mind* **74**: 79-83.

Stalnaker, R.C. 1970. 'Probability and Conditionals', *Philosophy of Science* **37**: 64-80.

Starke, P.H. 1969. *Abstracte Automaten.* Berlin: V.E.B. Deutscher Verlag der Wissenschaften.

Stevenson, C.L. 1970. 'If-iculties', *Philosophy of Science* **37**: 27-49.

Suppes, P. 1966. 'Probabilistic Inference and the Concept of Total Evidence', in *Aspects of Inductive Logic,* eds. J. Hintikka and P. Suppes, pp. 49-65. Amsterdam: North-Holland.

–, 1973a. 'Semantics of Context-Free Fragments of Natural Languages', in *Approaches to Natural Language,* eds. J. Hintikka et al., pp. 370-94. Dordrecht: D. Reidel.

–, 1973b. 'Comments on Montague's Paper', in *Approaches to Natural Language,* eds. J. Hintikka et al., pp. 259-62. Dordrecht: D. Reidel.

–, 1974. 'The Semantics of Children's Language', *American Psychologist* **29**: 103-14.

Tarski, A. 1956a. 'Fundamental Concepts of the Methodology of the Deductive Sciences', in *Logic, Semantics, Metamathematics,* ed. J.H. Woodger, pp. 60-109. London: Oxford University Press.

–, 1956b. 'The Concept of Truth in Formalized Languages', in *Logic, Semantics, Metamathematics,* eds. J.H. Woodger, pp. 152-278. London: Oxford University Press.

Tarski, A., Mostowski, A. and Robinson, R. 1953. *Undecidable Theories.* Amsterdam: North-Holland.

Teller, P. 1973. 'Conditionalization and Observation', *Synthese* **26**: 218-58.

Whorf, B.L. 1956. *Language, Thought and Reality: Selected Papers,* ed. J.B. Carroll. New York: Wiley.

Zadeh, L.A. 1965. 'Fuzzy Sets', *Information and Control* **8**: 338-53.

INDEX

GPSR Compliance

The European Union's (EU) General Product Safety Regulation (GPSR) is a set of rules that requires consumer products to be safe and our obligations to ensure this.

If you have any concerns about our products, you can contact us on ProductSafety@springernature.com

In case Publisher is established outside the EU, the EU authorized representative is:

Springer Nature Customer Service Center GmbH
Europaplatz 3
69115 Heidelberg, Germany

Batch number: 10830943

Printed by Printforce, the Netherlands